The Power of Geometric Algebra Computing

The Power of Geometric Algebra Computing

Dietmar Hildenbrand

CRC Press is an imprint of the
Taylor & Francis Group, an **informa** business
A CHAPMAN & HALL BOOK

First edition published 2022
by CRC Press
6000 Broken Sound Parkway NW, Suite 300, Boca Raton, FL 33487-2742

and by CRC Press
2 Park Square, Milton Park, Abingdon, Oxon, OX14 4RN

CRC Press is an imprint of Taylor & Francis Group, LLC

ISBN: 978-0-367-68458-7 (hbk)
ISBN: 978-0-367-68775-5 (pbk)
ISBN: 978-1-003-13900-3 (ebk)

DOI: 10.1201/9781003139003

Typeset in Nimbus Roman
by KnowledgeWorks Global Ltd.

Figure 1 Zeeland impression.

To Zeeland and all the wonderful memories and future experiences

Contents

List of Figures

List of Tables

Foreword

Geometric algebras have proved to be useful in engineering applications during the past three decades. It all began with Conformal Geometric Algebra (CGA) as a model of three-dimensional Euclidean space and its two-dimensional version, Compass Ruler Algebra (CRA). Both these outstanding algebras from the computing point of view were a topic for the previous two books by Dietmar Hildenbrand. It is only natural that his new book is treating newly reinvented structures, such as Projective Geometric Algebra (PGA), Geometric Algebra for Conics (GAC) or a Clifford algebra for quantum computing, Quantum Bit Algebra (QBA). The uniting element in all books by Dietmar Hildenbrand is GAALOP or its online version GAALOPWeb. Indeed, as the range of applications is growing, there is a need for a tool that is simply implementing symbolic operations in particular Clifford algebras into a code that can be easily included in standard programming languages or engineering software tools. Furthermore, if a geometric algorithm is precompiled by GAALOP, all operations are optimized with respect to the computational complexity. The advantage of such an approach lies also in the fact that the dimension of a Clifford algebra is not such a serious limitation regarding the number of operations and computational load. And finally, the geometric nature of the algorithms demands geometric imagination which has to be verified. Thus, GAALOPWeb provides a visualiszation tool based on a Python module, Ganja, where the geometric ideas may be checked instantly. Analogously to the previous books, even this one is easy to read because the author provides examples of applications for particular topics. Together with symbolic calculations, the reader finds code in the language for GAALOPWeb, which is GAALOPScript. Therefore, all applications may be simply checked, geometric primitives and operations calculated and visualized. Special focus is on quantum computing where GAALOPWeb serves also as a q-bit calculator. This is a novel concept which deserves attention and GAALOPWeb again proved its universality in accommodating different structures. After introducing the basics of geometric algebra computing for CGA, CRA and PGA, the book continues with the description of GAALOPWeb functionalities, including visualization. Consequently, four possibilities of generating a code for particular programming languages and engineering tools are described in their respective chapters. This includes C/C++, Python, Mathematica and MATLAB®, respectively. Each description is accompanied by specific engineering applications containing the code in GAALOPScript for instant verification in GAALOPWeb. Chapters 8 to 10 summarise the power of GAALOPWeb when used for high-dimensional geometric algebras, such as Double Conformal Geometric Algebra, Geometric Algebra for Conics and for Cubics. Subsequent two chapters deal with GAPP, the intermediate language of GAALOP and its implementation in GAALOPWeb. This is very suitable for hardware implementations; one of them is the Geometric Algebra hardware design GAPPCO I and, newly, GAPPCO II. Chapter 15 and the remaining chapters introduce the concept of quantum computing in a geometric algebra setting. Even in this area, GAALOPWeb proved its quality and is used as a q-bit calculator. As a summary, I believe that this book may well serve to support geometric algebras in penetrating engineering applications, yet it is suitable even in academia where it can guide both students and teachers in a fancy way. Indeed, the potential of geometric algebras may be exploited if and only if it can be introduced to a wide audience, and to do so, the calculations must be simple to implement, possible to be included in standard engineering software tools and visualized. And this is precisely what this book is about. It is my pleasure that the book will be introduced on the occasion of the 8th Conference on Applied Geometric Algebras in Computer Science and Engineering (AGACSE) in Brno, Czech Republic, 2021.

Prof. Petr Vasik
Brno University of Technology
Brno, Czech Republic
April 1, 2021

Preface

Geometric Algebra is a very powerful mathematical system for an easy and intuitive treatment of geometry, but the community working with it is still very small. The main goal of this book is to close this gap from a computing perspective in presenting the power of Geometric Algebra Computing for engineering applications and quantum computing. The intended audience includes students, engineers and researchers interested in really computing with Geometric Algebra.

Today, we indeed have the Geometric Algebra Computing technology available for easy-to-develop, geometrically intuitive, robust and fast engineering applications. We are happy to provide *GAALOPWeb*, the web-based version of our *GAALOP (GEOMETRIC ALGEBRA ALGORITHMS OPTIMIZER)* compiler for the integration of Geometric Algebra into standard programming languages. Besides the generation of optimized code for languages such as C++, Python, Mathematica, MATLAB and many others, GAALOPWeb can also be used for the configuration of a specific Geometric Algebra hardware, the GAPPCO[1] co-processor. It is also able to visualize Geometric Algebra algorithms. GAALOPWeb can be used easily on PC, smart phone, tablet etc. without any software installation.

While *The Power of Geometric Algebra Computing* really focuses on the easy-to-handle computing with Geometric Algebra, there is a following relation to my former books: *Foundation of Geometric Algebra Computing*[29] focuses on the GAALOP technology itself and on 3D applications for computer graphics, computer vision and robotics. The second book *Introduction to Geometric Algebra Computing*[30] is intended to give a rapid introduction to computing with Geometric Algebra. From the point of view of geometric objects, it focuses on the most basic ones, namely points, lines and circles. We call this algebra *Compass Ruler Algebra*, since you are able to handle it comparably to working with a compass and ruler. It offers the possibility to compute with these geometric objects, their geometric operations and transformations in a very intuitive way. While focusing on 2D, it is easily expandable to 3D computations as used in many books dealing with the popular Conformal Geometric Algebra.

The power of Geometric Algebra Computing can also be studied in this book by using high dimensional algebras for the handling of more complex geometric objects such as conics, quadrics and cyclides. Last, but not least, the advantageous use of Geometric Algebra for quantum computing is demonstrated.

[1]Geometric algorithms parallelization programs co-processor

I really do hope that this book can support the widespread use of Geometric Algebra as a mathematical tool for computing with geometry in engineering applications as well as in quantum computing.

Dr. Dietmar Hildenbrand

Acknowledegments

I would like to thank Dr.-Ing. Christian Steinmetz for his tremendous support to this book. He developed GAALOPWeb as used in this book during his PhD phase.

Many thanks to

- the group of Prof. Petr Vasik from Brno University of Technology for their support concerning the robot application, the Geometric Algebra of conics and quantum computing,

- Prof. Rafael Alves (Universidade Federal do ABC, Brasil) for his support regarding the molecular distance application as well as the application of Geometric Algebra to quantum computing,

- Dr. Silvia Franchini, Prof. A. Gentile, Prof. G. Vassallo and Prof. S. Vitabile from the University of Palermo for the cooperation regarding the new coprocessor design GAPPCO I,

- Robert Easter for the first integration of Double Conformal Geometric Algebra into GAALOP and his support with test scripts,

- Senior Associate Prof. Eckhard Hitzer for many fruitful discussions and for his big effort and enthusiasm for the promotion of Geometric Algebra (and computing with it).

CHAPTER 1

Introduction

CONTENTS

Geometric Algebra is a very powerful mathematical language combining geometric intuitivity with the potential of high runtime-performance of the implementations. This book on hand is based on *GAALOPWeb*, a new user-friendly, web-based tool for the generation of optimized code for different programming languages as well as for the visualization of Geometric Algebra algorithms for a wide range of engineering applications. It includes applications from the fields of computer graphics, robotics and quantum computing.

The book *Foundations of Geometric Algebra Computing* [29] describes GAALOP (see Chapt. 4) in a very fundamental way, since it breaks down the computing of Geometric Algebra algorithms into the most basic arithmetic operations. This book on hand makes use of its web-based extention *GAALOPWeb* (see Chapt. 4).

This book is suitable as a starting point for computing with Geometric Algebra for everybody interested in it as a new powerful mathematical system, especially for students, engineers and researchers in engineering, computer science, quantum computing and mathematics.

1.1 GEOMETRIC ALGEBRA

The main advantage of Geometric Algebra is its easy and intuitive treatment of geometry.

Geometric Algebra is based on the work of the German high school teacher Hermann Grassmann and his vision of a general mathematical language for geometry. His very fundamental work, called *Ausdehnungslehre*[24], was little noted in his time. Today, however, Grassmann is more and more respected as one of the most important mathematicians of the 19th century.

William Clifford [11] combined Grassmann's exterior algebra and Hamilton's quaternions in what we call *Clifford algebra* or *Geometric Algebra*[1]. Pioneering work

[1]David Hestenes writes in his article [27] about the genesis of Geometric Algebra: *Even today mathematicians typically typecast Clifford Algebra as the "algebra of a quadratic form," with no awareness*

DOI: 10.1201/9781003139003-1

has been done by David Hestenes. Especially interesting for this book is his work on Conformal Geometric Algebra (CGA) [26] [50].

1.2 GEOMETRIC ALGEBRA COMPUTING

Especially since the introduction of Conformal Geometric Algebra (see Sect. 2.2) there has been an increasing interest in using Geometric Algebra in engineering. The use of Geometric Algebra in engineering applications relies heavily on the availability of an appropriate computing technology. The main problem of *Geometric Algebra Computing* is the exponential growth of data and computations compared to linear algebra, since the **multivector**[2] of an n-dimensional Geometric Algebra is 2^n-dimensional. For the 5-dimensional Conformal Geometric Algebra, the multivector is already 32-dimensional.

An approach to overcome the runtime limitations of Geometric Algebra has been achieved through optimized software solutions. Tools have been developed for high-performance implementations of Geometric Algebra algorithms such as the C++ software library generator Gaigen 2 from Daniel Fontijne and Leo Dorst of the University of Amsterdam [21], GMac from Ahmad Hosney Awad Eid of Suez Canal University [20], the Versor library [12] from Pablo Colapinto and the C++ expression template library Gaalet [60] from Florian Seybold of the University of Stuttgart. BiVector.net (see Figure 1.1) gives a good overview over software solutions for

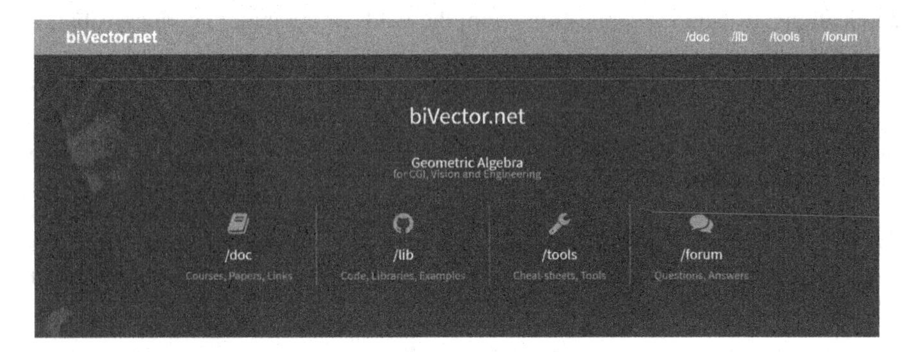

Figure 1.1 Screenshot of https://bivector.net.

Geometric Algebra. It provides code generators for C++, C#, Python, Rust as well as starting points for libraries for Python, C/C++, Julia and the JavaScript visualization tool Ganja (see Sect. 2.4).

Our GAALOP compiler [32] is not specific for one programming language but supports many of them. It can be used as a compiler for languages such as C/C++,

---

of its grander role in unifying geometry and algebra as envisaged by Clifford himself when he named it Geometric Algebra. It has been my privilege to pick up where Clifford left off — to serve, so to speak, as principal architect of Geometric Algebra and Calculus as a comprehensive mathematical language for physics, engineering and computer science.

[2]The main algebraic element of Geometric Algebra (please refer to Sect. 2.2)

C++ AMP, OpenCL and CUDA [29] [31] as well as Python, MATLAB, Mathematica,
Julia or Rust. Please find details about GAALOP in Chapt. 3 and about its web-based extension GAALOPWeb in Chapt. 4.

1.3 OUTLINE

Chapt. 2 presents the most important Geometric Algebras for engineering, namely Conformal Geometric Algebra (CGA), Compass Ruler Algebra (CRA) and Projective Geometric Algebra (PGA).

Chapt. 3 is dealing with *GAALOP* and **Chapt. 4** with *GAALOPWeb*, our new user-friendly, web-based version of GAALOP for a wide range of engineering applications based on Geometric Algebra algorithms. It makes it much easier for users to generate optimized source code without any software installation. GAALOPScript, the language to describe Geometric Algebra algorithms for the handling with GAALOPWeb is presented in Sect. 3.2. GAALOPWeb supports the user with visualizations of Geometric Algebra algorithms as demonstrated in Sect. 4.3. The visualization is based on Ganja as described in Sect. 2.4.

In the following chapters some applications of GAALOPWeb for different programming languages are presented. *GAALOPWeb for C/C++* is presented in **Chapt. 5** and *GAALOPWeb for Python* in **Chapt. 6**. A Molecular Distance Application using *GAALOPWeb for Mathematica* is shown in **Chapt. 7** and an application of robot kinematics based on *GAALOPWeb for MATLAB* in **Chapt. 8**.

The power of high-dimensional Geometric Algebras, as presented in **Chapt. 9**, comes with their ability to easily handle complex geometric objects. While the default Geometric Algebra of GAALOPWeb is the 5D Conformal Geometric Algebra with points, spheres, planes, circles and lines as geometric objects, three additional algebras are presented. The handling of conics in GAALOPWeb is shown in **Chapt. 10**. Quadrics as another example of geometric objects, which are interesting for applications, are handled in **Chapt. 11** and cubics in **Chapt. 12**.

Geometric Algebra has an inherent potential for parallelization. This can be very well seen in an intermediate language of GAALOP called GAPP. **Chapt. 13** describes GAALOPWeb for this language which is very well suitable for hardware implementations. **Chapt. 14** shows, how *GAALOPWeb for GAPPCO* is dealing with the Geometric Algebra hardware design GAPPCO. **Chapt. 15** describes a new hardware design called GAPPCO II.

Also quantum computing benefits a lot from Geometric Algebra. This is due to its ability to describe the quantum bit (qubit) operations as geometric transformations. The basics of quantum computing are presented in **Chapt. 16**. The principle of a new Geometric Algebra for qubits together with some examples is presented in **Chapt. 17**. We extended GAALOPWeb for it in a way to use it as a qubit calculator.

CHAPTER 2

Geometric Algebras for Engineering

CONTENTS

This chapter presents the basics of Geometric Algebra as well as the most important Geometric Algebras for engineering, namely Conformal Geometric Algebra (CGA), Compass Ruler Algebra (CRA) and Projective Geometric Algebra (PGA).

2.1 THE BASICS OF GEOMETRIC ALGEBRA

The main product of Geometric Algebra is called the **geometric product**; many other products can be derived from it. The three most often used products of Geometric Algebra are the **outer**, the **inner** and the **geometric** product. The notations of these products are listed in Table 2.1. We will use the outer product mainly for the construction and intersection of geometric objects, while the inner product will be used for the computation of angles and distances. The geometric product will be used mainly for the description of transformations.

Basis Blades are the elementary algebraic elements of Geometric Algebra. An n-dimensional Geometric Algebra consists of basis blades with **grades** 0 to n, where

DOI: 10.1201/9781003139003-2

TABLE 2.1 Notations of Geometric Algebra

Notation	Meaning
AB	Geometric product of A and B
$A \wedge B$	Outer product of A and B
$A \cdot B$	Inner product of A and B
A^{-1}	Inverse of A
$\tilde{A}$	Reverse of A
A^*	Dual of A

TABLE 2.2 Properties of the Outer Product $\wedge$

Property	Meaning
Anticommutativity	$u \wedge v = -(v \wedge u)$
Distributivity	$u \wedge (v + w) = u \wedge v + u \wedge w$
Associativity	$u \wedge (v \wedge w) = (u \wedge v) \wedge w$

a scalar is a **0-blade** (a blade of grade 0) and the **1-blades** are the basis vectors. The **2-blades** $e_i \wedge e_j$[1] are basis blades spanned by two 1-blades, and so on. There exists only one element of the maximum grade n. It is therefore also called the **pseudoscalar**. A linear combination of k-blades is called a k-vector (or a vector, bivector, trivector. ...). A linear combination of arbitrary blades is called a **multivector**. Multivectors are the general elements of a Geometric Algebra. The properties of the **outer product** are listed in Table 2.2. The first property applies only to vectors; the others are generally valid (and so they are also valid for multivectors). The outer product of two parallel vectors is 0:

$$u \wedge u = -(u \wedge u) = 0. \tag{2.1}$$

This is the reason why the outer product can be used as a measure of parallelness.

For Euclidean Geometric Algebras[2], the **inner product** $u \cdot v$ of two vectors is the same as the well-known Euclidean scalar product of two vectors. For perpendicular vectors, the inner product is 0; for instance, $e_1 \cdot e_2 = 0$. In Geometric Algebra, however, the inner product is not only defined for vectors. The inner product is grade-decreasing; for example, the result of the inner product of elements with grade 2 and 1 is an element of grade $2 - 1 = 1$. See Sect. 3.2.7 of [57] for a mathematical treatment of the general inner product.

The **geometric product** AB is an amazingly powerful operation, which is used mainly for the handling of transformations. The geometric product of vectors is a combination of the outer product and the inner product. The geometric product of u and v is denoted by uv. For vectors u and v, the geometric product uv can be defined

[1] the symbol $\wedge$ means the outer product of Geometric Algebra

[2] all basis vectors square to 1

as

$$uv = u \wedge v + u \cdot v. \tag{2.2}$$

We derive the following for the inner and outer products:

$$u \cdot v = \frac{1}{2}(uv + vu), \tag{2.3}$$

$$u \wedge v = \frac{1}{2}(uv - vu). \tag{2.4}$$

but, as noted above, these formulas apply only for vectors, in this form. See Chap. 3 of [57] for further details of the products of Geometric Algebra and especially Sect. 3.1 for an axiomatic approach to the geometric product.

The **inverse** A^{-1} of a blade A is defined by

$$AA^{-1} = 1.$$

The inverse of a vector v, for instance, is

$$v^{-1} = \frac{v}{v \cdot v}.$$

since

$$v\frac{v}{v \cdot v} = \frac{v \cdot v}{v \cdot v} = 1.$$

The inverse of the vector $v = 2e_1$ for instance results in $0.5e_1$, since $v \cdot v = 2$. See [39] for details about multivector inverses.

Since the geometric product is invertible, divisions by algebraic expressions are possible. The **dual** A^* of an algebraic expression is calculated by dividing it by the pseudoscalar I. In the following, the dual of the pseudoscalar $e_1 \wedge e_2$ is calculated. A superscript * means the dual operator.

$$(e_1 \wedge e_2)^* = (e_1 \wedge e_2)(e_1 \wedge e_2)^{-1}$$

$$(e_1 \wedge e_2)^* = (e_1 \wedge e_2)\underbrace{(e_1 \wedge e_2)^{-1}}_{-(e_1 \wedge e_2)}$$

$$(e_1 \wedge e_2)^* = -(e_1 \wedge e_2)(e_1 \wedge e_2)$$

$$(e_1 \wedge e_2)^* = -\underbrace{(e_1 \wedge e_2)(e_1 \wedge e_2)}_{-1}$$

$$(e_1 \wedge e_2)^* = 1.$$

See [57] for mathematical details.

The **reverse** $\tilde{A}$ of a multivector is the multivector with reversed order of the outer product components; for instance the reverse of $1 + e_1 \wedge e_2$ is $1 + e_2 \wedge e_1$ or $1 - e_1 \wedge e_2$.

2.2 CONFORMAL GEOMETRIC ALGEBRA (CGA)

Conformal Geometric Algebra is very well suitable to realize engineering applications. This is primarily because of its easy handling of geometric objects such as spheres, planes and lines.

Conformal Geometric Algebra uses three Euclidean basis vectors e_1, e_2, e_3 and two additional orthogonal basis vectors e_+, e_-

$$e_+ \cdot e_- = 0 \tag{2.5}$$

with positive and negative signatures, respectively, which means that they square to $+1$ as usual (e_+) and to -1 (e_-)

$$e_+^2 = 1, \qquad e_-^2 = -1. \tag{2.6}$$

Another basis e_0, e_∞, with the geometric meaning where

e_0 represents the 3D origin,

e_∞ represents infinity,

can be defined by the transformation

$$e_0 := \frac{1}{2}(e_- - e_+), \qquad e_\infty := e_- + e_+. \tag{2.7}$$

TABLE 2.3 The 32 basis blades of the 5D Conformal Geometric Algebra (Conformal Geometric Algebra)

Grade	Term	Blades	No.
0	scalar	1	1
1	basis vectors	$e_1, e_2, e_3, e_\infty, e_0$	5
2	2-blades (for bivectors)	$e_1 \wedge e_2,\ e_1 \wedge e_3,\ e_1 \wedge e_\infty,$ $e_1 \wedge e_0,\ e_2 \wedge e_3,\ e_2 \wedge e_\infty,$ $e_2 \wedge e_0,\ e_3 \wedge e_\infty,\ e_3 \wedge e_0,$ $e_\infty \wedge e_0$	10
3	3-blades (for trivectors)	$e_1 \wedge e_2 \wedge e_3,\ e_1 \wedge e_2 \wedge e_\infty,\ e_1 \wedge e_2 \wedge e_0,$ $e_1 \wedge e_3 \wedge e_\infty,\ e_1 \wedge e_3 \wedge e_0,\ e_1 \wedge e_\infty \wedge e_0,$ $e_2 \wedge e_3 \wedge e_\infty,\ e_2 \wedge e_3 \wedge e_0,\ e_2 \wedge e_\infty \wedge e_0,$ $e_3 \wedge e_\infty \wedge e_0$	10
4	4-blades (for quadvectors)	$e_1 \wedge e_2 \wedge e_3 \wedge e_\infty,$ $e_1 \wedge e_2 \wedge e_3 \wedge e_0,$ $e_1 \wedge e_2 \wedge e_\infty \wedge e_0,$ $e_1 \wedge e_3 \wedge e_\infty \wedge e_0,$ $e_2 \wedge e_3 \wedge e_\infty \wedge e_0$	5
5	pseudoscalar	$e_1 \wedge e_2 \wedge e_3 \wedge e_\infty \wedge e_0$	1

Table 2.3 shows the 32 basis blades of Conformal Geometric Algebra, consisting of the scalar, five (basis) vectors, ten 2-blades, ten 3-blades, five 4-blades and

the pseudoscalar. A linear combination of k-blades is called a k-vector (or a vector, bivector, trivector. ...). The sum $e_2 \wedge e_3 + e_1 \wedge e_2$, for instance, is a bivector. A linear combination of arbitrary blades is called a multivector as for instance the sum $3 + e_1 \wedge e_2$.

2.2.1 Geometric Objects of Conformal Geometric Algebra

Conformal Geometric Algebra provides a great variety of basic geometric entities to compute with, namely points, spheres, planes, circles, lines, and point pairs, as listed in Table 2.4. These entities have two algebraic representations: the IPNS (inner product null space) and the OPNS (outer product null space). The IPNS of the algebraic expression A are all the points X satisfying the equation

$$A \cdot X = 0. \tag{2.8}$$

The OPNS of the algebraic expression A are all the points X satisfying the equation

$$A \wedge X = 0. \tag{2.9}$$

These representations are duals of each other (a superscript asterisk denotes the dualization operator). In Table 2.4, the vectors $\mathbf{x}$ and $\mathbf{n}$ are in bold type to indicate that they represent 3D entities obtained by linear combinations of the 3D basis vectors e_1, e_2, and e_3:

$$\mathbf{x} = x_1 e_1 + x_2 e_2 + x_3 e_3. \tag{2.10}$$

In the OPNS representation, the outer product $\wedge$ indicates the construction of a geometric object with the help of points P_i that lie on it. A sphere, for instance, is defined by four points ($P_1 \wedge P_2 \wedge P_3 \wedge P_4$) on its surface. In the IPNS representation, the meaning of the outer product is an intersection of geometric entities. A circle, for instance, is defined by the intersection of two spheres $S_1 \wedge S_2$. In order to compute it according to Fig. 2.1, you first have to compute the centre points P_1, P_2

TABLE 2.4 The two representations (IPNS and OPNS) of conformal geometric entities. They are duals of each other, which is indicated by the asterisk symbol.

Entity	**IPNS representation**	**OPNS representation**
Point	$P = \mathbf{x} + \frac{1}{2}\mathbf{x}^2 e_\infty + e_0$	
Sphere	$S = P - \frac{1}{2}r^2 e_\infty$	$S^* = P_1 \wedge P_2 \wedge P_3 \wedge P_4$
Plane	$\pi = \mathbf{n} + d e_\infty$	$\pi^* = P_1 \wedge P_2 \wedge P_3 \wedge e_\infty$
Circle	$C = S_1 \wedge S_2$	$C^* = P_1 \wedge P_2 \wedge P_3$
Line	$L = \pi_1 \wedge \pi_2$	$L^* = P_1 \wedge P_2 \wedge e_\infty$
Point pair	$Pp = S_1 \wedge S_2 \wedge S_3$	$Pp^* = P_1 \wedge P_2$

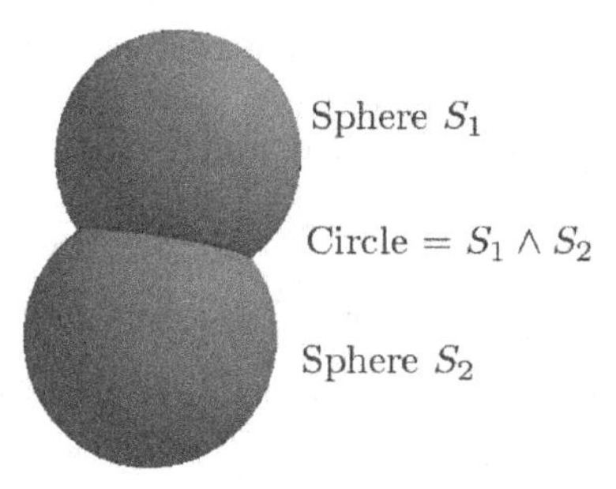

Figure 2.1 The intersection of two spheres results in a circle.

$$P_1 = \mathbf{x_1} + \frac{1}{2}\mathbf{x_1}^2 e_\infty + e_0,$$
$$P_2 = \mathbf{x_2} + \frac{1}{2}\mathbf{x_2}^2 e_\infty + e_0,$$

based on their 3D representations $\mathbf{x_1}, \mathbf{x_2}$

$$\mathbf{x_1} = x_{1,1}e_1 + x_{2,1}e_2 + x_{3,1}e_3 \tag{2.11}$$

and

$$\mathbf{x_2} = x_{1,2}e_1 + x_{2,2}e_2 + x_{3,2}e_3 \tag{2.12}$$

and then compute the spheres S_1, S_2 with these points and the radii r_1 and r_2

$$S_1 = P_1 - \frac{1}{2}r_1^2 e_\infty,$$
$$S_2 = P_2 - \frac{1}{2}r_2^2 e_\infty,$$

and finally compute the circle as the intersection of them using the outer product

$$Circle = S_1 \wedge S_2.$$

2.2.2 Angles and Distances in 3D

The inner product of geometric objects have the meaning of angles or distances in Conformal Geometric Algebra according to Table 2.5. Details can be found in the indicated sections of [29].

2.2.3 3D Transformations

Reflections are the basis of 3D transformations. The reflection of an object o at a plane P is expressed by

$$o_{\text{reflected}} = PoP. \tag{2.13}$$

TABLE 2.5 Geometric Meaning of the Inner Product of Planes, Spheres and Points

·	Plane	Sphere	Point
Plane	Angle between planes [29] Sect 3.4.2	Euclidean distance from centre [29] Sect 3.4.1.3	Euclidean distance [29] Sect 3.4.1.2
Sphere	Euclidean distance from centre [29] Sect 3.4.1.3	Distance measure [29] Sect 3.4.1.4	Distance measure [29] Sect 3.4.1.5
Point	Euclidean distance [29] Sect 3.4.1.2	Distance measure [29] Sect 3.4.1.5	Distance [29] Sect 3.4.1.1

Transformations such as rotations and translations are also represented by combinations of reflections. The operator

$$R = e^{-(\frac{\phi}{2})L} \tag{2.14}$$

describes a **rotor**. L is the rotation axis, represented by a normalized bivector, and ϕ is the rotation angle around this axis. R can also be written as

$$R = \cos\left(\frac{\phi}{2}\right) - L\sin\left(\frac{\phi}{2}\right). \tag{2.15}$$

The rotation of a geometric object o is performed with the help of the operation

$$o_{\text{rotated}} = Ro\tilde{R}.$$

There are strong relations between rotations in Conformal Geometric Algebra and quaternions and dual quaternions (see [29]).

In Conformal Geometric Algebra, a translation can be expressed in a multiplicative way with the help of a **translator** T defined by

$$T = e^{-\frac{1}{2}\mathbf{t}e_\infty}, \tag{2.16}$$

where $\mathbf{t}$ is a vector

$$\mathbf{t} = t_1e_1 + t_2e_2 + t_3e_3.$$

Application of the Taylor series

$$T = e^{-\frac{1}{2}\mathbf{t}e_\infty} = 1 + \frac{-\frac{1}{2}\mathbf{t}e_\infty}{1!} + \frac{(-\frac{1}{2}\mathbf{t}e_\infty)^2}{2!} + \frac{(-\frac{1}{2}\mathbf{t}e_\infty)^3}{3!} + \ldots$$

and the property $(e_\infty)^2 = 0$ results in the translator

$$T = 1 - \frac{1}{2}\mathbf{t}e_\infty. \tag{2.17}$$

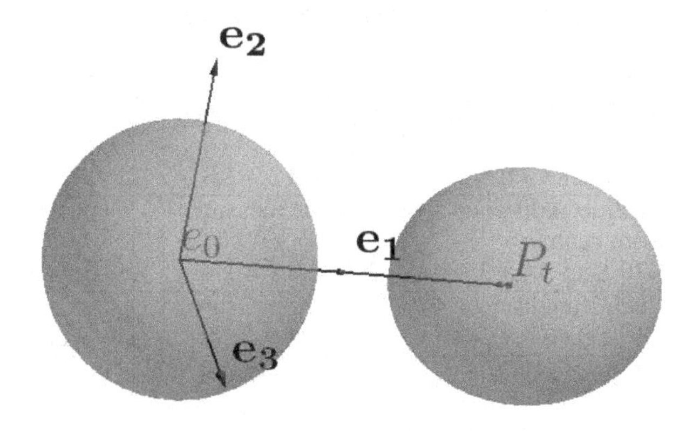

Figure 2.2 Translation of a sphere from the origin to the point P_t.

Let us, for instance, translate the sphere

$$S = -e_\infty + e_0 \tag{2.18}$$

(see Fig. 2.2) in the x-direction by the translation vector

$$\mathbf{t} = 4e_1. \tag{2.19}$$

Note that this is a sphere with its centre at the origin and with $r^2 = 2$. The translator in this example has the form

$$T = 1 - 2e_1e_\infty, \tag{2.20}$$

and its reverse is

$$\tilde{T} = 1 + 2e_1e_\infty. \tag{2.21}$$

The translated sphere can now be computed as the versor product

$$S_{\text{translated}} = TS\tilde{T} \tag{2.22}$$

$$= (1-2e_1e_\infty)(-e_\infty + e_0)(1+2e_1e_\infty)$$

$$= (1-2e_1e_\infty)(-e_\infty - 2\underbrace{e_\infty e_1 e_\infty}_{0} + e_0 + 2e_0e_1e_\infty)$$

$$= (1-2e_1e_\infty)(-e_\infty + e_0 - 2e_1e_0e_\infty)$$

$$= -e_\infty + e_0 - 2e_1e_0e_\infty + 2e_1\underbrace{e_\infty e_\infty}_{0} - 2e_1e_\infty e_0 + 4e_1e_\infty e_1e_0e_\infty$$

$$= -e_\infty + e_0 - 2e_1\underbrace{(e_0e_\infty + e_\infty e_0)}_{-2} + 4e_1e_\infty e_1e_0e_\infty$$

$$= 4e_1 - e_\infty + e_0 + 4\underbrace{e_1e_\infty e_1}_{-e_\infty}e_0e_\infty$$

$$= 4e_1 - e_\infty + e_0 - 4e_\infty \underbrace{e_0 e_\infty}_{e_0 \wedge e_\infty - 1}$$

$$= 4e_1 - e_\infty + e_0 - \underbrace{4e_\infty(e_0 \wedge e_\infty)}_{-e_\infty} + 4e_\infty,$$

resulting in

$$S_{\text{translated}} = 4e_1 + 7e_\infty + e_0. \tag{2.23}$$

This is a sphere with the same radius $r^2 = 2$, but with a translated centre point

$$P_t = \mathbf{t} + \frac{1}{2}\mathbf{t}^2 e_\infty + e_0 = 4e_1 + 8e_\infty + e_0. \tag{2.24}$$

Please notice that in 3D a rigid body motion is still more general in the sense that it consists of a rotation around an arbitrary line in space together with a translation in the direction of this line. Please refer to Chapt. 8 for an application of a robot kinematics using these transformations. For more details about Conformal Geometric Algebra see [29].

2.3 COMPASS RULER ALGEBRA (CRA)

Compass Ruler Algebra [30] is simply the Conformal Geometric Algebra in 2D. This 4D Geometric Algebra[3] is represented by multivectors with 16 coordinates, namely combinations of the four basis vectors $(e_1, e_2, e_0, e_\infty)$.

As follows, we will see how

- geometric objects and their intersections
- angles and distances between geometric objects
- transformations of geometric objects

can be expressed easily with the help of algebraic expressions.

2.3.1 Geometric Objects

Table 2.7 shows a list of the basic geometric objects of the Compass Ruler Algebra, namely points, circles, lines and point pairs.

These entities have two algebraic representations, the IPNS and the OPNS representation. These representations are duals of each other (a superscript asterisk denotes the dualization operator). A 2D point with coefficients x_1, x_2 and basis vectors e_1, e_2

$$\mathbf{x} = x_1 e_1 + x_2 e_2 \tag{2.25}$$

is embedded in the 4D Compass Ruler Algebra as point

$$P = \mathbf{x} + \frac{1}{2}\mathbf{x}^2 e_\infty + e_0 \tag{2.26}$$

[3] 4D relates to the number of four basis vectors

TABLE 2.6 The 16 Basis Blades of the Compass Ruler Algebra (to be identified by their indices)

Index	Blade
0	1
1	e_1
2	e_2
3	e_∞
4	e_0
5	$e_1 \wedge e_2$
6	$e_1 \wedge e_\infty$
7	$e_1 \wedge e_0$
8	$e_2 \wedge e_\infty$
9	$e_2 \wedge e_0$
10	$e_\infty \wedge e_0$
11	$e_1 \wedge e_2 \wedge e_\infty$
12	$e_1 \wedge e_2 \wedge e_0$
13	$e_1 \wedge e_\infty \wedge e_0$
14	$e_2 \wedge e_\infty \wedge e_0$
15	$e_1 \wedge e_2 \wedge e_\infty \wedge e_0$

with the two additional basis vectors e_∞, e_0 (with the geometric meaning of infinity and origin) and

$$\mathbf{x}^2 = x_1^2 + x_2^2 \tag{2.27}$$

being the scalar product of $\mathbf{x}$.

$\mathbf{x}$ and $\mathbf{n}$ in Table 2.7 are in bold type to indicate that they represent 2D entities obtained by linear combinations of the 2D basis vectors e_1 and e_2. L represents a line with normal vector $\mathbf{n}$ and distance d to the origin. The $\{Ci\}$ represents different circles. The outer product "$\wedge$" indicates the construction of a geometric object with the help of points $\{P_i\}$ that lie on it. A circle, for instance, is defined by three points $(P_1 \wedge P_2 \wedge P_3)$ on this circle. Another meaning of the outer product is the intersection of geometric entities. A point pair is defined by the intersection of two circles $C_1 \wedge C_2$.

The following example shows how, in Compass Ruler Algebra, we are able to compute comparable to working with compass and ruler. In order to construct the perpendicular bisector of a section of a line with compass and ruler, we draw two circles with the centre at the boundary points and connect the two intersection points according to Fig. 2.3.

TABLE 2.7 The Representations of the Geometric Objects of the Compass Ruler Algebra

Entity	IPNS representation	OPNS representation
Point	$P = \mathbf{x} + \frac{1}{2}\mathbf{x}^2 e_\infty + e_0$	
Circle	$C = P - \frac{1}{2}r^2 e_\infty$	$C^* = P_1 \wedge P_2 \wedge P_3$
Line	$L = \mathbf{n} + d e_\infty$	$L^* = P_1 \wedge P_2 \wedge e_\infty$
Point pair	$P_p = C_1 \wedge C_2$	$P_p^* = P_1 \wedge P_2$

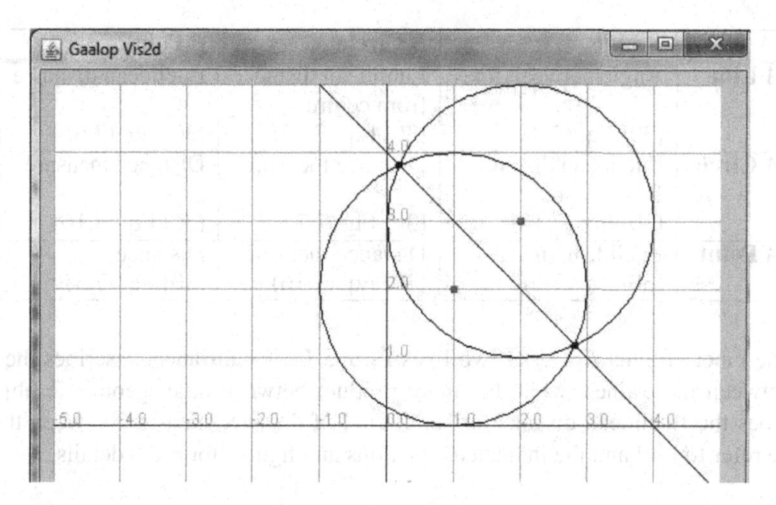

Figure 2.3 Vis2d Visualization of the perpendicular bisector between the two (red) points.

At first we have to compute the circles based on the centre points P_1, P_2 and the radii r_1, r_2[4]

$$S_1 = P_1 - \frac{1}{2}r_1^2 e_\infty,$$

$$S_2 = P_2 - \frac{1}{2}r_2^2 e_\infty,$$

and then compute the intersection of them

$$PP = S_1 \wedge S_2$$

based on the outer product. The bisector is then the line through the two points of the resulting point pair

$$L = (PP^* \wedge e_\infty)^*$$

[4]Please notice that the first part of the algorithm is completely the same as for spheres in 3D according to the algorithm of Sect. 2.2.1

Sect. 3.2.3 shows how to describe this algorithm in GAALOP and Sect. 4.3.1 how to visualize it. Please refer to Sect. 5.2 for some runtime considerations based on this example.

2.3.2 Angles and Distances

The inner product of these geometric objects describes distances and angles between them as summarized in Table 2.8.

TABLE 2.8 Geometric Meaning of the Inner Product of Lines, Circles and Points

$A \cdot B$	*B* **Line**	*B* **Circle**	*B* **Point**
A **Line**	Angle between lines [30] Eq. (7.9)	Euclidean distance from centre [30] Eq. (7.13)	Euclidean distance [30] Eq. (7.6)
A **Circle**	Euclidean distance from centre [30] Eq. (7.13)	Distance measure [30] Fig. 7.7	Distance measure [30] Eq. (7.16)
A **Point**	Euclidean distance [30] Eq. (7.6)	Distance measure [30] Eq. (7.16)	Distance [30] Eq. (7.3)

The inner product $L_1 \cdot L_2$ of two lines L_1 and L_2, for instance, describes the angle between these lines, while the inner product between other geometric objects describes the Euclidean distance or some kind of distance measure between them. Please refer to [30] and the indicated equations and figures for more details.

2.3.3 Transformations

Transformations of a geometric object o can be easily described within Compass Ruler Algebra according to Table 2.9. The reflection, for instance, of a circle C at a line L can be computed based on the (geometric) product $-LCL$ (please remember that the geometric product in Geometric Algebra is written without a specific product symbol). Rotations or translations can be described based on algebraic expressions

TABLE 2.9 The Description of Transformations of a Geometric Object o in Compass Ruler Algebra (please note that LoL means the geometric product of L, o and L)

Transformation	**Operator**	**Usage**
Reflection	$L = \mathbf{n} + de_\infty$	$o_L = -LoL$
Rotation	$R = \cos\left(\frac{\phi}{2}\right) - e_1 \wedge e_2 \sin\left(\frac{\phi}{2}\right)$	$o_R = Ro\tilde{R}$
Translation	$T = 1 - \frac{1}{2}\mathbf{t}e_\infty$	$o_T = To\tilde{T}$

called rotors R and translators T. Using the rotor

$$R = \cos\left(\frac{\phi}{2}\right) - e_1 \wedge e_2 \sin\left(\frac{\phi}{2}\right), \tag{2.28}$$

the rotated object o_R can be computed based on the geometric product $Ro\tilde{R}$

$$o_R = Ro\tilde{R} \tag{2.29}$$

with $\tilde{R}$ being the *reverse* of R (see Sect. 2.1). A translated object o_T can be computed based on the translator

$$T = 1 - \frac{1}{2}\mathbf{t}e_\infty \tag{2.30}$$

with $\mathbf{t}$ being the 2D translation vector $t_1e_1 + t_2e_2$ as

$$o_T = To\tilde{T}. \tag{2.31}$$

Please refer to [30] for more details.

2.4 PROJECTIVE GEOMETRIC ALGEBRA (PGA) WITH GANJA

The Projective Geometric Algebra (PGA) has been introduced to a wider computer graphics community at the Siggraph conference 2019 by Charles Gunn and Steven de Keninck. Ganja [14] is a javascript-based package for the web, developed by Steven de Keninck. It allows the computation of Geometric Algebra algorithms, which are written in javascript, and provides also a visualization of the Geometric Algebra algorithms. The Ganja Coffeeshop [15] is equipped with some examples of PGA visualizations using Ganja[14].

2.4.1 2D PGA

PGA is presented on some cheat sheets [13] at www.bivector.net. Figure 2.4 shows some basics of 2D PGA. It is a Geometric Algebra with signature (2,0,1) means there are two basis vectors squaring to 1, none squaring to -1 and one squaring to 0. The three basis vectors are the vectors e_0 squaring to 0 and the basis vectors e_1 and e_2 squaring to 1. Additionally, 2D PGA consists of the scalar, the three bivectors e_{01}, e_{20} and e_{12} and the pseudoscalar e_{012}. Because of e_0 squaring to 0, the multiplication table of these basis blades consists of many zero values.

Figure 2.5 presents some geometry basics of 2D PGA. The Euclidean point (x,y) is computed as a bivector sum and a line as a vector sum. There are also some incidence, reject and project operations shown. Meeting of two geometric objects means intersecting them and joining of two points, for instance, means constructing a line based on these points.

Listing 2.1 shows a simple Ganja script from the Ganja Coffeeshop. It computes with points and lines by creating, joining and intersecting them. The result is shown in Fig. 2.6.

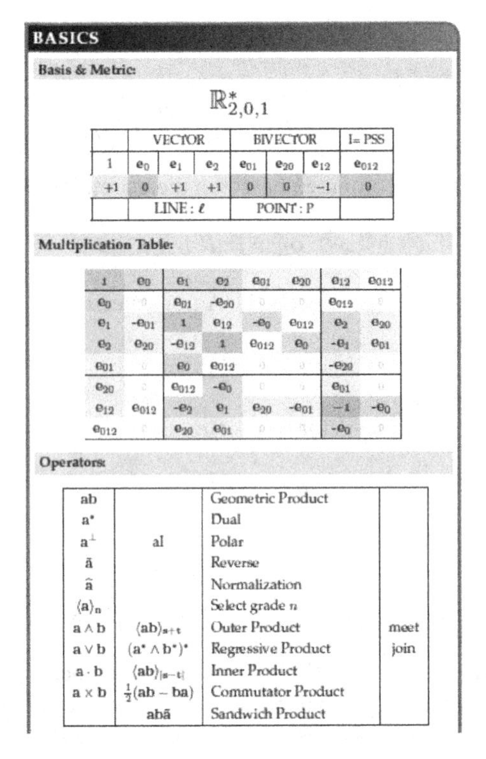

BASICS

Basis & Metric:

$$\mathbb{R}^*_{2,0,1}$$

	VECTOR			BIVECTOR			I= PSS
1	e_0	e_1	e_2	e_{01}	e_{20}	e_{12}	e_{012}
+1	0	+1	+1	0	0	−1	0
	LINE : ℓ			POINT : P			

Multiplication Table:

1	e_0	e_1	e_2	e_{01}	e_{20}	e_{12}	e_{012}
e_0	0	e_{01}	$-e_{20}$	0	0	e_{012}	0
e_1	$-e_{01}$	1	e_{12}	$-e_0$	e_{012}	e_2	e_{20}
e_2	e_{20}	$-e_{12}$	1	e_{012}	e_0	$-e_1$	e_{01}
e_{01}	0	e_0	e_{012}	0	0	$-e_{20}$	0
e_{20}	0	e_{012}	$-e_0$	0	0	e_{01}	0
e_{12}	e_{012}	$-e_2$	e_1	e_{20}	$-e_{01}$	−1	$-e_0$
e_{012}	0	e_{20}	e_{01}	0	0	$-e_0$	0

Operators:

ab		Geometric Product	
a^*		Dual	
$a^\perp$	aI	Polar	
$\tilde{a}$		Reverse	
$\hat{a}$		Normalization	
$\langle a\rangle_n$		Select grade n	
$a \wedge b$	$\langle ab\rangle_{s+t}$	Outer Product	meet
$a \vee b$	$(a^* \wedge b^*)^*$	Regressive Product	join
$a \cdot b$	$\langle ab\rangle_{\lvert s-t\rvert}$	Inner Product	
$a \times b$	$\frac{1}{2}(ab - ba)$	Commutator Product	
	$ab\tilde{a}$	Sandwich Product	

Figure 2.4 The basics of the 2D Projective Geometric Algebra according to [13].

Listing 2.1: simple Ganja script according to [15].

```
Algebra(2,0,1,()=>{

    var point = (x,y)=>1e12-x*1e02+y*1e01;
    var line = (a,b,c)=>a*1e1+b*1e2+c*1e0;

    var A=point(-1,1), B=point(-1,-1), C=point(1,-1);
    var L=line(-1,1,0.5)
    var M=()=>C&A;
    var D=()=>L^M;

    document.body.appendChild(this.graph([
        A, "A",             // Render point A and label it.
        B, "B",             // Render point B and label it.
        C, "C",             // Render point C and label them.
        L, "L", M, "M",     // Render and label lines.
        D, "D",             // Intersection point of L and M
        0xff0000,           // Set the color to red.
        [B,C],              // Render line segment from B to C.
        0xffcccc,           // Set the color to light red.
        [A,B,C]             // render polygon ABC.
    ],{grid:true}));
});
```

GEOMETRY	
Points, Lines:	
Euclidean point at (x, y)	$x\mathbf{e}_{20} + y\mathbf{e}_{01} + \mathbf{e}_{12}$
Direction (ideal point) (x, y)	$x\mathbf{e}_{20} + y\mathbf{e}_{01}$
Line with eq. $ax + by + c = 0$	$\boldsymbol{\ell} = a\mathbf{e}_1 + b\mathbf{e}_2 + c\mathbf{e}_0$
Incidence:	
Join points $\mathbf{P}_1, \mathbf{P}_2$ in line $\boldsymbol{\ell}$	$\boldsymbol{\ell} = \mathbf{P}_1 \vee \mathbf{P}_2$
Meet lines $\boldsymbol{\ell}_1, \boldsymbol{\ell}_2$ in point P	$P = \boldsymbol{\ell}_1 \wedge \boldsymbol{\ell}_2$
Project, Reject:	
Line orthogonal to line $\boldsymbol{\ell}$, through point P	$\boldsymbol{\ell} \cdot \mathbf{P} = \boldsymbol{\ell} \times \mathbf{P}$
Project point P on line $\boldsymbol{\ell}$	$(\boldsymbol{\ell} \cdot \mathbf{P})\boldsymbol{\ell}$
Project line $\boldsymbol{\ell}$ on point P	$(\boldsymbol{\ell} \cdot \mathbf{P})\mathbf{P}$
Direction orthogonal to line $\boldsymbol{\ell}$	$\boldsymbol{\ell}^{\perp} := \boldsymbol{\ell}I$

Figure 2.5 Geometry basics of the 2D Projective Geometric Algebra according to [13].

First, we create a Geometric Algebra with 2,0,1 metric, means two basis vectors are squaring to 1 and one basis vector squaring to 0. We define our points to be bivectors[5]. For readability we create a function that converts 2D Euclidean coordinates to their bivector representation. Similarly, we can define lines directly. The Euclidean line ax + by + c can be specified as a*1e1+b*1e2+c*1e0[6].

We then define the 3 points A, B and C. They are based on their 2D coordinates but draggable in the visualization. The line L is defined according to the equation $y = x - 0.5$. A line can also be defined by joining[7] two points. We define M as a function so it will update when C or A are dragged. Points can also be found by intersecting (meeting) two lines. We similarly define D as a function for interactive updates. Finally, we use the graph function to create the visualization of our algebraic elements (see Fig. 2.6). The graph function accepts an array of items that it will render in order. It can render points, lines, labels, colours, line segments and polygons.

2.4.2 3D PGA

3D PGA is also described by a cheat sheet [13] at www.bivector.net. It is a natural extension of 2D PGA as can be seen on Fig. 2.7 with the geometry basics of 3D PGA.

[5]please take notice of the minus sign of x because of the anti-commutativity of the outer product
[6]in Ganja, the basis vectors or basis blades are written with a number 1 in front
[7]Please refer to the definition in Fig. 2.4

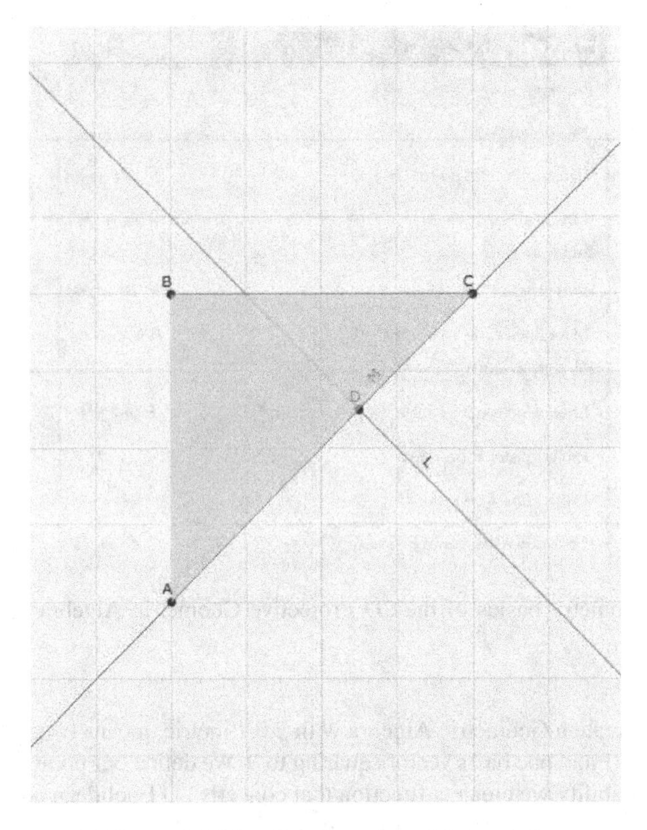

Figure 2.6 Visualization of the Listing 2.1.

The following example of the Ganja Coffeeshop according to Listing 2.2 handles transformations as motion interpolations between geometric objects.

Listing 2.2: Ganja script for 3D transformations according to [15].

```
1  // Create a Geometric Algebra with 3,0,1 metric.
2  Algebra(3,0,1,()=>{
3
4    // The geometric elements of 3D PGA.
5    // Grade-1 elements (reflections) represent planes.
6    // Grade-2 elements (bireflections or line reflections)
7    // represent lines.
8    // Grade-3 elements (trireflections or point reflections)
9    // represent points.
10   var plane = (a,b,c,d)
11         =>(a*1e1 + b*1e2 + c*1e3 + d*1e0).Normalized,
12       line  = (...plucker)
13         =>(plucker*[1e01,1e02,1e03,1e12,1e13,1e23]).Normalized,
14       point = (x,y,z)=>!(1e0 + x*1e1 + y*1e2 + z*1e3);
15
16   // Transformations in PGA are represented by versors,
17   // which are compositions (geometric product)
```

GEOMETRY	
Points, Lines, Planes :	
Euclidean point (x, y, z)	$\mathbf{P} = x\mathbf{e}_{032} + y\mathbf{e}_{013} + z\mathbf{e}_{021} + \mathbf{e}_{123}$
Ideal point (direction) (x, y, z)	$\mathbf{P} = x\mathbf{e}_{032} + y\mathbf{e}_{013} + z\mathbf{e}_{021}$
Plane $ax + by + cz + d = 0$	$\mathbf{p} = a\mathbf{e}_1 + b\mathbf{e}_2 + c\mathbf{e}_3 + d\mathbf{e}_0$
Incidence: (5)	
Join points/directions $\mathbf{P}_1, \mathbf{P}_2$ in line $\boldsymbol{\ell}$	$\boldsymbol{\ell} = \mathbf{P}_1 \vee \mathbf{P}_2$
Meet planes $\mathbf{p}_1, \mathbf{p}_2$ in line $\boldsymbol{\ell}$	$\boldsymbol{\ell} = \mathbf{p}_1 \wedge \mathbf{p}_2$
Join points $\mathbf{P}_1, \mathbf{P}_2, \mathbf{P}_3$ in plane $\mathbf{p}$	$\mathbf{p} = \mathbf{P}_1 \vee \mathbf{P}_2 \vee \mathbf{P}_3$
Meet planes $\mathbf{p}_1, \mathbf{p}_2, \mathbf{p}_3$ in point $\mathbf{P}$	$\mathbf{P} = \mathbf{p}_1 \wedge \mathbf{p}_2 \wedge \mathbf{p}_3$
Join line $\boldsymbol{\ell}$ and point $\mathbf{P}$ in plane $\mathbf{p}$	$\mathbf{p} = \boldsymbol{\ell} \vee \mathbf{P}$
Meet line $\boldsymbol{\ell}$ and plane $\mathbf{p}$ in point $\mathbf{P}$	$\mathbf{P} = \boldsymbol{\ell} \wedge \mathbf{p}$
Project, Reject:	
Plane $\perp$ to plane $\mathbf{p}$ through line $\boldsymbol{\ell}$	$\mathbf{p} \cdot \boldsymbol{\ell}$
Line $\perp$ to plane $\mathbf{p}$ through point $\mathbf{P}$	$\mathbf{p} \cdot \mathbf{P}$
Plane $\perp$ to line $\boldsymbol{\ell}$ through point $\mathbf{P}$	$\boldsymbol{\ell} \cdot \mathbf{P}$
Project plane $\mathbf{p}$ onto point $\mathbf{P}$ (6)	$(\mathbf{p} \cdot \mathbf{P})\mathbf{P}$
Project point $\mathbf{P}$ onto plane $\mathbf{p}$	$(\mathbf{p} \cdot \mathbf{P})\mathbf{p}$
Project plane $\mathbf{p}$ onto line $\boldsymbol{\ell}$	$(\mathbf{p} \cdot \boldsymbol{\ell})\boldsymbol{\ell}$
Project line $\boldsymbol{\ell}$ onto plane $\mathbf{p}$	$(\mathbf{p} \cdot \boldsymbol{\ell})\mathbf{p}$
Project line $\boldsymbol{\ell}$ onto point $\mathbf{P}$	$(\boldsymbol{\ell} \cdot \mathbf{P})\mathbf{P}$
Project point $\mathbf{P}$ onto line $\boldsymbol{\ell}$	$(\boldsymbol{\ell} \cdot \mathbf{P})\boldsymbol{\ell}$
Direction $\perp$ to plane $\mathbf{p}$	$\mathbf{p}^{\perp} = \mathbf{p}\mathbf{I}$
Ideal line $\perp$ to line $\boldsymbol{\ell}$	$\boldsymbol{\ell}^{\perp} = \boldsymbol{\ell}\mathbf{I}$

Figure 2.7 Geometry basics of the 3D Projective Geometric Algebra according to [13].

```
18  // of reflections (vectors).
19
20  // Both rotations and translations can be generated
21  // by exponentiating
22  var motor = (line,angle_or_distance)
23       =>Math.E**(angle_or_distance/2 * line);
24
25  // As in 2D finding transforms between elements
26  // only requires the square root.
27  var sqrt = motor => ((Math.sign(motor.s) + motor)
28       *(1 + 0.5*(motor.s + motor.Grade(4)))).Normalized;
29
30  // motors can be interpolated
31  // (similar to quaternions in two ways).
32  var lerp = (motor, x) => (Math.sign(motor().s)*(1-x)
33       + x*motor).Normalized;
34
35  // Create some points (P), lines (l), planes (p) to illustrate.
```

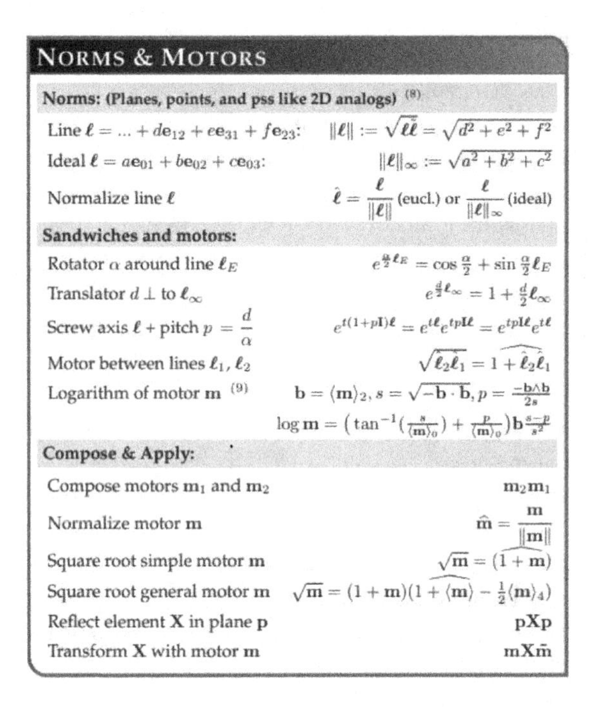

Figure 2.8 Norms and motors of the 3D Projective Geometric Algebra according to [13].

```
36    var P1 = point(0,0,-1),        P2 = point(.3,0.4,-1),
37        p1 = plane(0,1,0,0.2),     p2 = plane(-0.5,0.5,0,.5),
38        l1 = line(0,0,.5,.1,1,0), l2 = line(0,0,.7,-.1,1,0);
39
40    // Motors between all these elements :
41    var P1toP2 = ()=>sqrt(P2 * P1),
42        p1top2 = ()=>sqrt(p2 * p1),
43        l1tol2 = ()=>sqrt(l2 * l1);
44
45    // Graph it
46    document.body.appendChild(this.graph(()=>{
47      // Animation time.
48      var t = 0.5+0.5*Math.sin(performance.now()/1000);
49
50      // A rotation around the z-axis.
51      var R = motor(line(0,0,0,1,0,0),t*Math.PI*2);
52
53      return [
54        "Rotations␣and␣Translations",
55        0x882288,lerp(p1top2, t) >>> p1*0.3, // move between planes
56        0x882288,lerp(l1tol2,t) >>> l1, // move between lines
57        0x882288,lerp(P1toP2,t) >>> P1, // move between points.
58        0x224488,[P1,R >>> P2],"R", // R applied to P2
59        0x008844,P1,"P1",P2,"P2",  // our two points.
60        0x008844,p1*0.3,"p1",p2*0.3,"p2", // our two planes.
```

```
61          0x228844,l1,"l1",l2,"l2", // our two lines.
62        ];
63    },{grid:true, labels:true,lineWidth:3, h:-0.4, p:-0.1,
64          scale:1.4, animate:true}));
65
66  });
```

At first, the Geometric Algebra is created with 3 basis vectors squaring to 1 and 1 basis vector squaring to 0. After the definition of the motor functionality and lerp as the motion interpolator, the objects to be interpolated are defined. The visualization according to Fig. 2.9 shows the green points, lines and planes together with the animated magenta ones interpolating them.

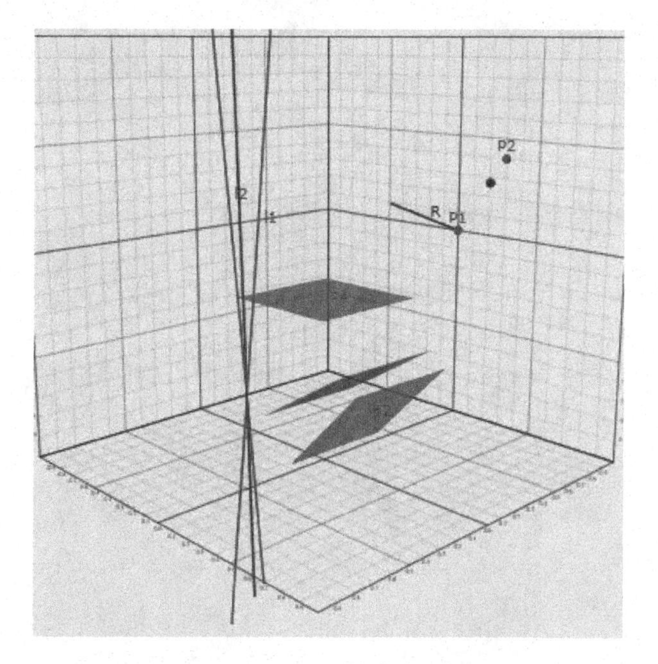

Figure 2.9 Visualization of the Listing 2.2.

Ganja is not only available for PGA but also for arbitrary Geometric Algebras. Later on we are using Ganja for the visualizations of GAALOPWeb (see Sect. 4.3).

CHAPTER 3

GAALOP

CONTENTS

The **G**eometric **A**lgebra **al**gorithms **op**timizer GAALOP is a free software tool in order to optimize Geometric Algebra algorithms (download via [33]). A good way of cutting the high complexity of Geometric Algebra before going to the real computing device is to precompute / precompile Geometric Algebra algorithms with GAALOP as described in the book "Foundations of Geometric Algebra Computing" [29]. The big potential of optimizations of Geometric Algebra algorithms before runtime can be very good demonstrated with the inverse kinematics algorithm of [34] [28], which was in 2006 the first Geometric Algebra application that was faster than the standard implementation. In this application more than 99% of computing time can be saved and less than 1% is left to the computing at runtime. In this computer animation application, the Geometric Algebra implementation based on symbolic simplification was three times faster than the conventional solution.

In 2009, a very remarkable result in terms of runtime performance could be achieved with the robot grasping algorithm of [66]. Based on optimized C-code generation with GAALOP, a speedup of 14 could be achieved compared to the conventional mathematics solution. In the meantime, GAALOP has been extended to support many programming languages. It can be used as a compiler for languages such as C/C++, C++ AMP, OpenCL and CUDA [29] [31] as well as Python, MATLAB and Mathematica.

To be a bit more precise, GAALOP takes a Geometric Algebra algorithm in the form of a *GAALOPScript* (see Sect. 3.2) and generates functions for different programming languages, which are optimized in terms of high runtime-performance and numerical stability. GAALOPScripts are sequential programs as a list of assignments of multivectors to Geometric Algebra computations. Principally, GAALOP is performing a symbolic precomputation of the coefficients of the multivectors

DOI: 10.1201/9781003139003-3

resulting in very simple elementary operations (simple additions and multiplications). In a nutshell, GAALOP takes an algorithm on the level of Geometric Algebra operations, computes the resulting multivectors symbolically and optimizes the calculations of each coefficient by symbolic simplifications.

3.1 INSTALLATION

The stand-alone GAALOP as a JAVA program can be downloaded free of charge from *http://www.gaalop.de*. We recommend also to install Maxima [58] in order to be able to use the complete optimization potential of GAALOP. Fig. 3.1 shows how

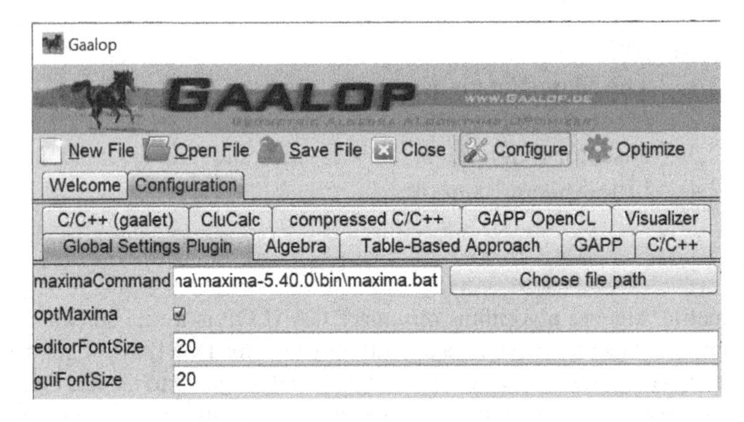

Figure 3.1 Global Setting Plugin for the configuration of Maxima (as well as font sizes).

GAALOP has to be configured for the use of Maxima. In the Global Setting Plugin the path of the file *maxima.bat* of the Maxima installation has to be chosen and *optMaxima* has to be activated.

As an example we describe the configuration for Compass Ruler Algebra (see Sect. 2.3), since this algebra is used in the book "Introduction to Geometric Algebra Computing" [30] with a recommended tutorial (Chapt. 3 of [30]). The screenshot in Fig. 3.2 shows how GAALOP should be configured for the use of Compass Ruler

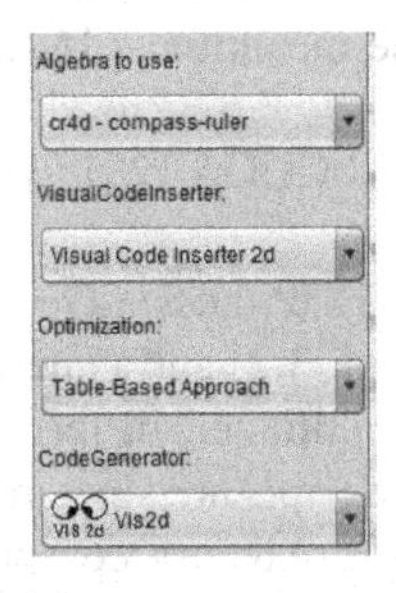

Figure 3.2 Configuration of GAALOP for visualizations based on Compass Ruler Algebra.

Algebra. Please select the "cr4d - compass-ruler" as "Algebra to use" and the "Visual Code Inserter 2d" for 2d visualizations to be performed by "Vis2d", the "CodeGenerator" to be selected for the purpose of visualization. Please also select the default "Table-Based Approach" for "Optimization".

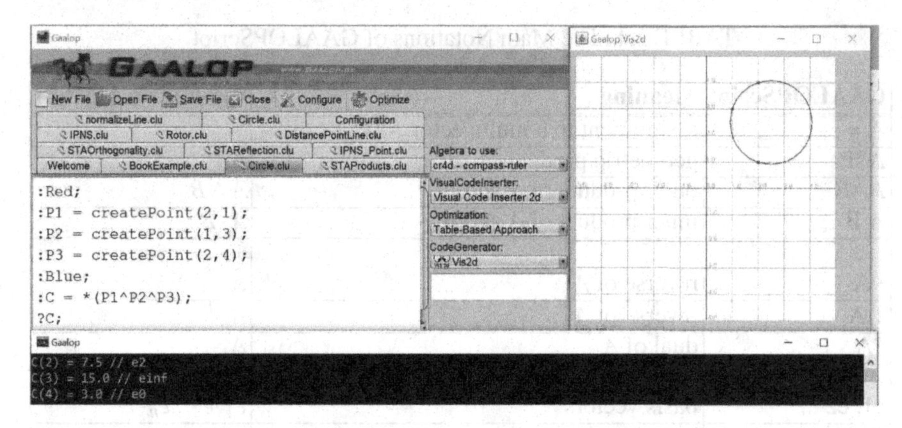

Figure 3.3 Screenshot of the editor, the visualization and the output window of GAALOP.

The screenshot in Fig. 3.3 shows the three windows of GAALOP. The editor window is responsible for the administration and for the editing of *GAALOPScripts* (the input language of GAALOP according to Sect. 3.2). Aside is the CodeGenerator/visualization window. If Vis2d is selected for 2D visualizations of GAALOPScripts, the output window (at the bottom) is able to show numeric values of multivectors.

Here is the current list of all the Geometric Algebras of GAALOP with short and long name:

```
2d;imaginary numbers
3d;vectors in 3d
cra;compass-ruler
sta;space-time
cga;conformal space
gac;conformal conic
dcga;double conformal
qga;GA by Zamora
cuga;cubics
```

The definition of new algebras and their visualization is described in Chapt. 9.

3.2 GAALOPScript

GAALOPScript is a language for the description of Geometric Algebra algorithms. GAALOPScript is derived from CLUScript [57] and somewhat similar to the C programming language. Just as in C, every program line has to be ended by a semicolon.

The advantage of this convention is that a program line can be extended over a number of text lines as well as one program line can consist of several short commands. Comments can be included in the script in the same way as in C (see Table 3.1). Please find some information about GAALOPScript in the following sections.

TABLE 3.1 The Main Notations of GAALOPScript

GAALOPScript	Meaning	Geometric Algebra
A = ...	assignment to a multivector	
A*B	geometric product	AB
A∧B	outer product of A and B	$A \wedge B$
A.B	inner product of A and B	$A \cdot B$
~A	reverse of A	$\tilde{A}$
1/A	inverse of A	A^{-1}
*A	dual of A	A^*
e1, e2 ...	basis vectors	$e_1, e_2 ... e_n$
// /* .. */	comment in one line comment over lines	
?A	explicitely compute multivector A Note: at least one mulitvector of a GAALOPScript has to be marked this way	
!A	compute multivector A but only the coefficients which are needed for further computations	
:A	visualize multivector A	
:Color(r,g,b)	set colour to the defined RGB values (with values between 0 and 1), predefined colours are: :Black :Blue :Cyan :Green :Magenta :Orange :Red :White :Yellow	

3.2.1 The Main Notations

Table 3.1 summarizes the most important notations of GAALOPScript and Table 3.2 shows some notations specific for Conformal Geometric Algebra and Compass Ruler Algebra.

TABLE 3.2 Notations of GAALOPScript for CGA/CRA

GAALOPScript	Meaning	CGA/CRA
e1, e2	2D basis vectors	e_1, e_2
e3	3rd basis vector specific for CGA	e_3
e0	origin	e_0
einf	infinity	e_∞

One essential part of GAALOPScripts are assignments of Geometric Algebra computations to multivectors. Please notice that while for the geometric product no specific symbol is used, in GAALOPScript "*" is needed as a symbol. The operators for the dual and the reverse of multivector A are written in front of A. The inverse of A is expressed as "1/A". As an example, the following listing shows a simple GAALOPScript for the computation of the geometric product of two vectors.

Listing 3.1: GAALOPScript for the geometric product of two vectors.

```
1 a = a1*e1 + a2*e2 + a3*e3;
2 b = b1*e1 + b2*e2 + b3*e3;
3 ?c = a*b;
```

On the left side of an assignment, there is either a scalar variable or a multivector variable[1] while unknown variables are always assumed to be scalars. In our example, a, b and c are multivectors while *a1*, *a2* and *a3* as well as *b1*, *b2* and *b3* are scalar values.

First, the two multivectors a and b are computed as a linear combination of the 3D basis vectors. In the last line, the multivector c is computed as the geometric product of them. The question mark in front of a multivector indicates that we are interested in the explicit computation of this multivector. This means, that we are here interested in the result of c, while a and b are only intermediate results. Please notice that at least one multivector of a GAALOPScript has to be marked in this way.

A colon in front of a multivector means that this multivector should be visualized. Colours for the visualizations can be defined either by predefined colours or by a predefined function based on RGB (red, green, blue) values. Please refer to Sect. 4.3 for details about visualizations based on GAALOPScript.

[1] multiple assignments to variables are not allowed

TABLE 3.3 Macros of GAALOPScript for Conformal Geometric Algebra

Macro	Meaning
createPoint(x,y,z)	Creates a conformal point with the coordinates x, y, z
Sphere(centre, radius)	Creates a sphere from a given centre and a given radius Note: centre is a conformal point
Sphere(cx, cy, cz, radius)	Creates a sphere from given centre coordinates and a given radius
Rotor(x,y,z,angle)	Creates a rotor, which rotates along an axis (defined by x,y,z) with an angle
Translator(x,y,z)	Creates a translator, which translates along a vector (defined by x,y,z)
ExtractFirstPoint(pp)	Extracts the first point of a given point pair
ExtractSecondPoint(pp)	Extracts the second point of a given point pair
Dual(mv)	Dualizes a given multivector
Normalize(mv)	Normalizes a given multivector mv

3.2.2 Macros and Pragmas

There are predefined macros in order to simplify the development of GAALOP-Scripts. Table 3.3 shows a list of GAALOPScript macros for Conformal Geometric Algebra, for instance for the generation of spheres, rotors or translators, for extracting the two points of point pairs ... The user is able to write own macros using the following structure:

```
MacroName = {
  optional GAALOPScript
  ReturnExpression
}
```

The multivector to be returned is defined in the last line of the macro. Optionally, GAALOPScript assignments can be defined. The parameters are denoted by `_P(1)`, `_P(2)`, ... Please find examples, for instance, in the Sects. 8.4 and 9.1.

There are pragmas for the control of the generation of optimized code according to Table 3.4. Please find an example for the use of the *output* pragma in Sect. 4.3.4. The *range* pragma is used for sliders according to Sec. 4.3.5.

The pragma *in2out* can be used in order to define the sequence of the function parameters (otherwise lexical order is used) or if multivectors should only be intermediate results which should not be returned.

TABLE 3.4 Pragmas of GAALOPScript

Pragma	Meaning
output	If a question mark is set at the specified variable, only the specified blades are calculated, Ex.; //#pragma output mv 1.0 e3 e1 ∧ e2
in2out	defines the input variables and the multivectors to be returned. Ex.: //#pragma in2out x1, x2, x3 -> mv1, mv2
range	defines the range of a variable for visualizations for visualizations. //#pragma range minValue<=VariableName<=maxValue Ex.: //#pragma range 1<=a1<= 5

3.2.3 Bisector Example

The bisector example of Sect. 2.3.1 uses Compass Ruler Algebra and can be expressed in GAALOPScript as follows:

Listing 3.2: GAALOPScript for the bisector example.

```
1  P1 = createPoint(x1,y1);
2  P2 = createPoint(x2,y2);
3
4  S1 = P1 - 0.5*r*r*einf;
5  S2 = P2 - 0.5*r*r*einf;
6
7  PP = S1^S2;
8
9  ?L = *(*PP^einf);
```

It computes the perpendicular bisector of a line segment defined by the two points $P1$ and $P2$. The macro "createPoint" creates a point based on the two coordinates of a point in $\mathbb{R}^2$. The symbol $\wedge$ stands for the outer product, and $*$ before some element, like in the expression for L, represents the dualization. The expressions for $S1$ and $S2$ represent circles, and L is the bisector of these spheres. It is important to remark that only L has a question mark, this indicates that only the array L will be the result. Section 4.3.1 shows how to visualize this example and Sect. 5.2 presents the generated C/C++ code together with some runtime considerations.

3.2.4 Line-Sphere Example

The papers [31] and [25] present an example of a ray tracing application. One part of the corresponding algorithm is the computation whether a ray is intersectiong a (bounding) sphere or not. This can be expressed by the following GAALOPScript[2]

Listing 3.3: GAALOPScript for the computation whether there is an intersection of a ray with a sphere or not.

[2]here, all geometric objects are described in the IPNS representation according to Table 2.4

```
1 S = createPoint(Cx, Cy, Cz) - 0.5*r*r*einf;
2 O = createPoint(Ox, Oy, Oz);
3 L = createPoint(Lx, Ly, Lz);
4 R = *(O ^ L ^ einf);
5 PP = R ^ S;
6 ?hasIntersection = PP.PP;
```

This example (as also used in Chapt. 6) computes the sphere S and the ray R through the points O and L (createPoint computes a conformal point based on its 3D coordinates). The intersection of the sphere S and the ray R can easily be expressed with the help of the outer product of these two geometric entities (Fig. 3.4). The

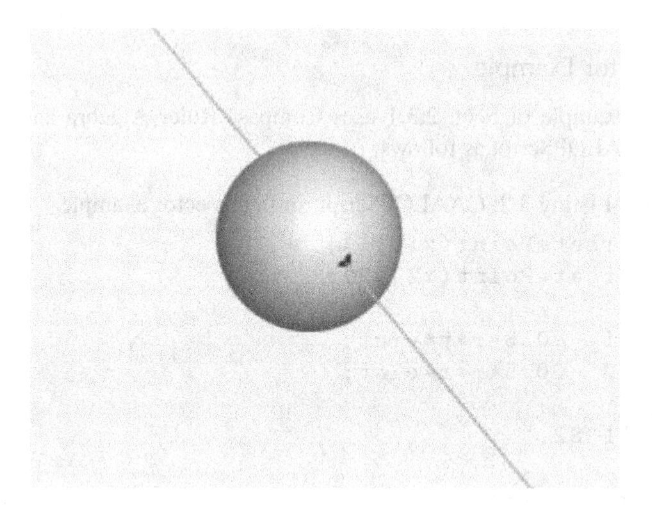

Figure 3.4 Spheres and lines are basic entities of Geometric Algebra that one can compute with. Intersection of these objects are easily expressed with the help of their outer product. Here, one of the two points of the intersection of the sphere and the line is shown in blue.

result is the point pair PP. In the script we call its norm *hasIntersection*, since its sign indicates whether the ray and the sphere are really intersecting each other or not. The question mark at the beginning of this line indicates that this multivector is the only one of this script to be computed explicitly[3].

Please refer to Sect. 4.3.3 for the visualization of this example.

[3]GAALOP is able to optimize not only single statements, but a number of Geometric Algebra statements. In the above script, the expressions for S, R, O, L and PP are used only by GAALOP, in order to compute an optimized result for the intersection indicator (see the question mark in the last line of the script)

CHAPTER 4

GAALOPWeb

CONTENTS

Here, we present GAALOPWeb[1], the new web-based GAALOP version. It can be run in every modern web browser and can therefore be used easily on PCs, smart phones, tablets, etc. without any software installation. GAALOPWeb uses Ganja (see Sect. 2.4) for the visualization of the Geometric Algebra algorithms. Figure 4.1 shows the selection of programming languages for the generation of optimized code. Currently C/C++, Julia, Mathematica, MATLAB, Python and Rust are supported.

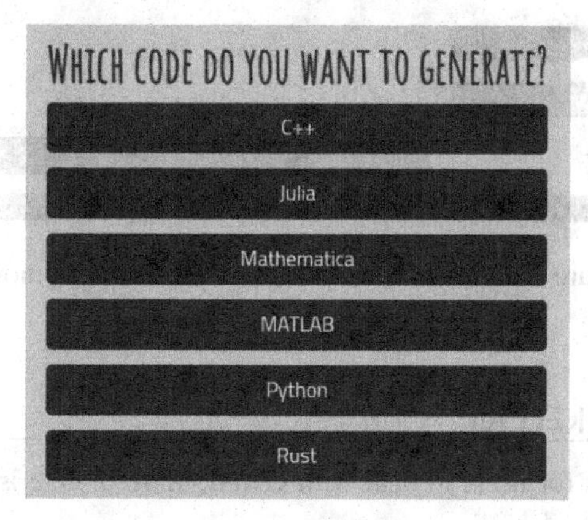

Figure 4.1 The programming language selection of GAALOPWeb.

[1] www.gaalop.de/gaalopweb/

DOI: 10.1201/9781003139003-4

4.1 THE WEB INTERFACE

Figure 4.2 shows the screenshot of *GAALOPWeb* with its main features:

- Conformal Geometric Algebra as default Geometric Algebra

- text area for the editing of GAALOPScripts for the *code to optimize*

- for the visualization of algorithms, there are additional text areas for *variable assignments* and the *multivectors to be visualized*

- there is a "Run" button in order to generate optimized code for the selected programming language and/or a visualization

- for supporting the user a quick help, some sample GAALOPScripts and an introduction to GAALOPScript is provided.

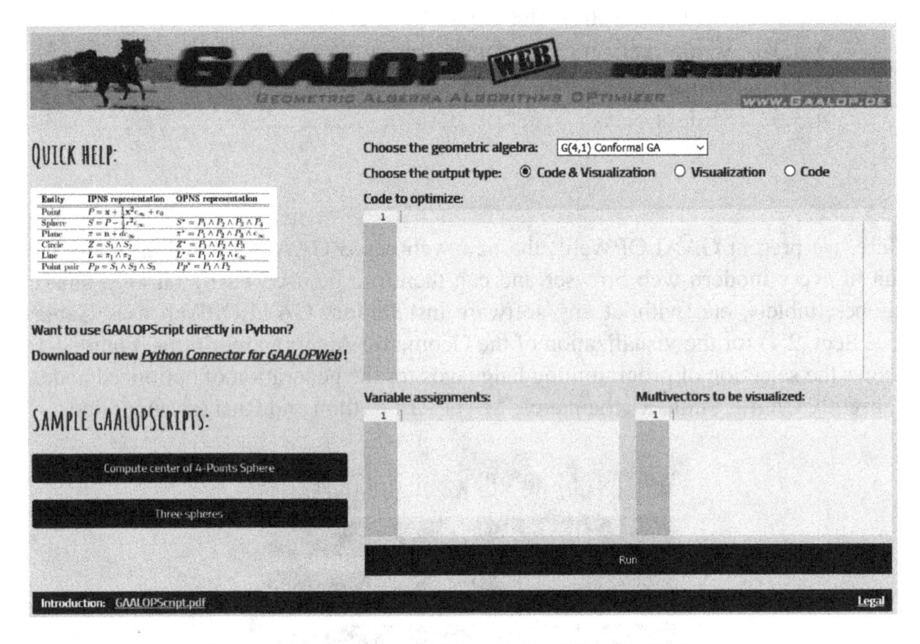

Figure 4.2 The screen of GAALOPWeb for Python.

4.2 THE WORKFLOW

There are two ways for users to deal with GAALOPWeb. One is simply using the web interface of Sect. 4.1

- edit your algorithm based on GAALOPScript

- optionally use the visualization (for the verification of your algorithm)

- generate code for the selected programming language

- use the optimized code in your environment based on copy and paste.

For users who would like to have a solution contained within their programming environment there is a second way:

- download the corresponding *Connector For GAALOPWeb* from the web page (see Fig. 4.2 for the download of the programming language connector for Python),

- edit a GAALOPScript (with extension .clu) as a text file and

- use the provided functionality in order to automatically generate an optimized function

  – for Python please refer to Chapt. 6,

  – for MATLAB please refer to Chapt. 8.

4.3 GAALOPWeb VISUALIZATIONS

A GAALOPScript including visualization is defined by three parts

1. Variable assignments

2. Code to optimize

3. Multivectors to be visualized

GAALOPWeb offers separate text fields for each of these parts (see the web interface in Sect. 4.1). Table 3.1 shows the GAALOPScript notations for visualizations, which are used in the following examples.

4.3.1 Visualization of the Bisector Example

Listing 4.1 shows the code for the visualization of the bisector example of Sect. 3.2.3.

Listing 4.1: GAALOPScript for the visualization of the bisector example.

```
1  // Variable assignments
2  x1=0.5; y1=0;
3  x2=1; y2=0.5;
4  r=1;
5
6  // Code to optimize
7  P1 = createPoint(x1,y1);
8  P2 = createPoint(x2,y2);
9
10 S1 = P1 - 0.5*r*r*einf;
```

```
11  S2 = P2 - 0.5*r*r*einf;
12
13  PP = S1^S2;
14
15  ?L = *(*PP^einf);
16
17  // Multivectors to be visualized
18  :Red;
19  :P1;
20  :P2;
21  :Blue;
22  :S1;
23  :S2;
24  :Black;
25  :L;
```

In addition to the known code to optimize we have to assign concrete values for the values x1, y1, x2, y2 and r. Furthermore we define the points P1, P2, the circles S1, S2 and the line L to be visualized together with their colours as can be seen in Fig. 4.3.

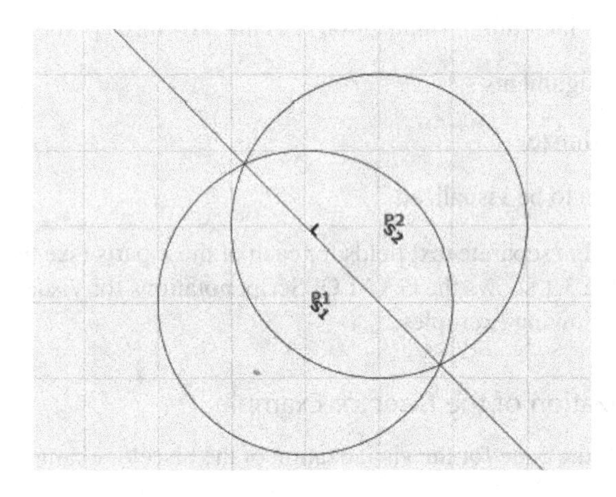

Figure 4.3 GAALOPWeb visualization of the bisector example.

This corresponds to the visualization of Sect. 3.2.5 of [30] for the stand-alone GAALOP. Please notice that all the visualizations of [30] can alternatively be done this way with GAALOPWeb.

4.3.2 Visualization of the Rotation of a Circle

Listing 4.2 shows the code for a rotation of a circle about 90 degrees around the origin according to Table 2.9.

Listing 4.2: GAALOPScript for the visualization of the bisector example.

```
1  // Variable assignments
2  x=1.5;
3  y=0;
4  r=0.5;
5  angle=90;
6
7  // Code to optimize
8  alpha=(angle/180)*3.1416;
9
10 P = createPoint(x,y);
11 Circle = P -0.5*r*r*einf;
12
13 Rota = cos(alpha/2) - (e1^e2)* sin(alpha/2);
14 ?Circle_rot = Rota * Circle * ~Rota;
15
16 // Multivectors to be visualized
17 :Red;
18 :Circle;
19 :Blue;
20 :Circle_rot;
21 :Black;
22 :Y=e1;
23 :X=e2;
```

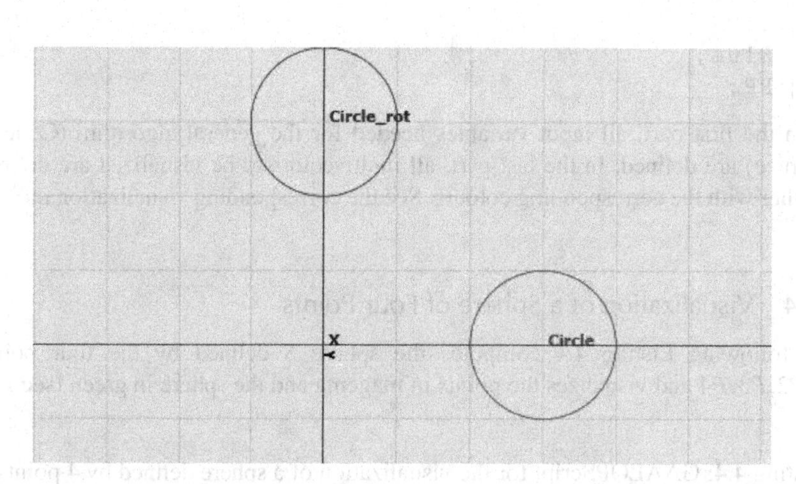

Figure 4.4 GAALOPWeb visualization of the rotation of a circle.

This corresponds to the visualization of Sect. 3.4.2 of [30] for the stand-alone GAALOP.

4.3.3 Visualization of the Line-Sphere Example

The Line-Sphere example of Sect. 3.2.4 can be visualized according to the following listing:

Listing 4.3: GAALOPScript for the visualization of an intersection of a ray with a sphere.

```
1  // Variable assignments
2  Cx =1;   Cy=1; Cz=1;
3  r=0.4;
4  Ox =1;   Oy=1.1; Oz=1;
5  Lx =1;   Ly=0.5; Lz=0.5;
6
7  // Code to optimize
8  S = createPoint(Cx, Cy, Cz) - 0.5*r*r*einf;
9  O = createPoint(Ox, Oy, Oz);
10 L = createPoint(Lx, Ly, Lz);
11 R = *(O ^ L ^ einf);
12 PP = R ^ S;
13 ?hasIntersection = PP.PP;
14
15 // Multivectors to be visualized
16 :Green;
17 :R;
18 :Yellow;
19 :S;
20 :Blue;
21 :PP;
```

In the first part, all input variables needed for the general algorithm (Code to optimize) are defined. In the last part, all multivectors to be visualized are defined together with the corresponding colours. See the corresponding visualization in Sect. 6.1.

4.3.4 Visualization of a Sphere of Four Points

The following Listing 4.4 computes the sphere S defined by the four points $P1, P2, P3, P4$ and visualizes the points in magenta and the sphere in green (see Fig. 4.5).

Listing 4.4: GAALOPScript for the visualization of a sphere defined by 4 points.

```
1  // Variable assignments
2  P1x=0; P1y=0; P1z=1;
3  P2x=0; P2y=1; P2z=0;
4  P3x=1; P3y=0; P3z=0;
5  P4x=0; P4y=0.707; P4z=0.707;
```

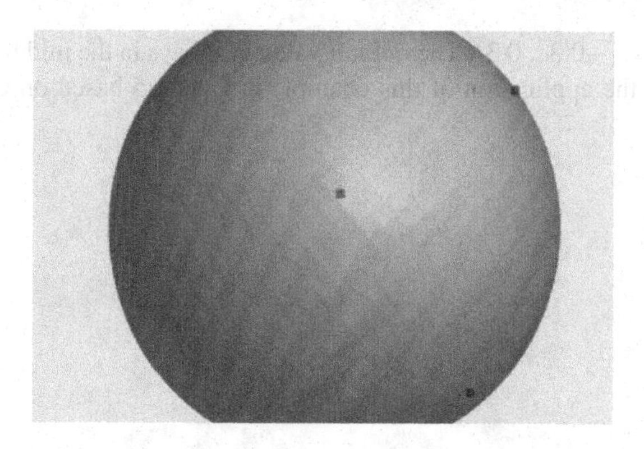

Figure 4.5 Visualization of the Listing 4.4 with four points defining a sphere.

```
6
7   // Code to optimize
8   P1 =  createPoint(P1x, P1y, P1z);
9   P2 =  createPoint(P2x, P2y, P2z);
10  P3 =  createPoint(P3x, P3y, P3z);
11  P4 =  createPoint(P4x, P4y, P4z);
12
13  S= P1^P2^P3^P4;
14
15  // Multivectors to be visualized
16  :Magenta;
17  :P1;
18  :P2;
19  :P3;
20  :P4;
21  :Green;
22  :S;
```

The sphere S is computed based on the outer product of the 4 points according to Table 2.4.

4.3.5 Sliders

In case that a variable is not defined for visualization, a slider is offered to adjust its value. The default range is [0 .. 1]. For a specific range the corresponding pragma can be used.

```
//#pragma range -0.3<=P1x<=0.3
P1y=0; P1z=1;
P2x=0; P2y=1; P2z=0;
P3x=1; P3y=0; P3z=0;
P4x=0; P4y=0.707; P4z=0.707;
```

Here the range is $[-0.3 .. 0.3]$. The default value is always in the middle of the range.

Please find the application of this example in Chapt. 5 based on GAALOPWeb for C.

CHAPTER 5

GAALOPWeb for C/C++

CONTENTS

C/C++ originally was the first language supported by GAALOP [29, 30]. There is also a precompiler available, called GAALOP GPC[1]. It is also available for GPUs as precompiler for OpenCL and CUDA[2] as well as for C++ AMP (please refer to [31]).

5.1 GAALOPWeb HANDLING

C/C++ users simply deal with GAALOPWeb using the web interface of Sect. 4.1. First of all, you have to select C++ as programming language. Then you are able to edit your algorithm based on GAALOPScript. Fig. 5.1 shows the screen for the computation of a sphere defined by four points.

[1]please refer to Chapter 11 of [29]
[2]please refer to Chapter 12 of [29]

DOI: 10.1201/9781003139003-5

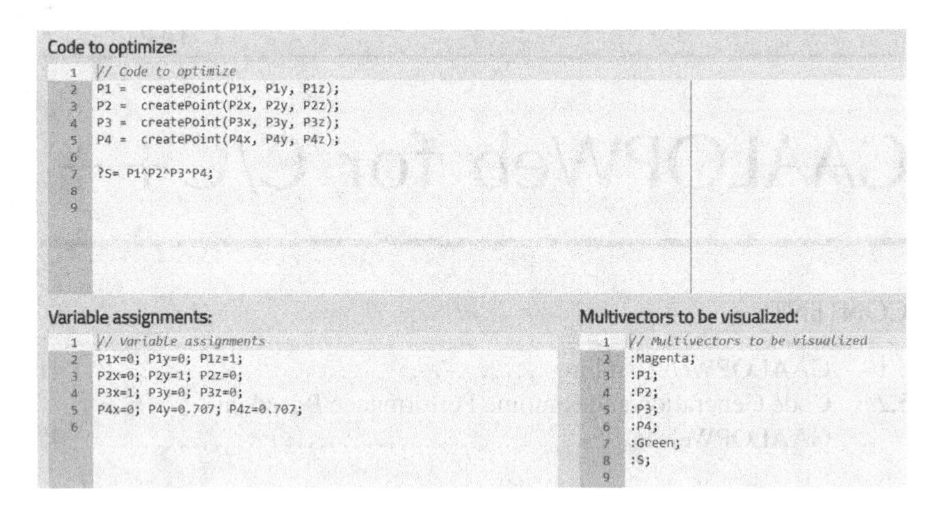

Figure 5.1 The sphere-centre example with GAALOPWeb.

The algorithm based on Sect. 4.3.4 can be seen in the *Code to optimize* area. The information for the visualization is defined in the *Variable assignments* and the *Multivectors to be visualized* area. Fig. 5.2 shows the screen for the results of the example.

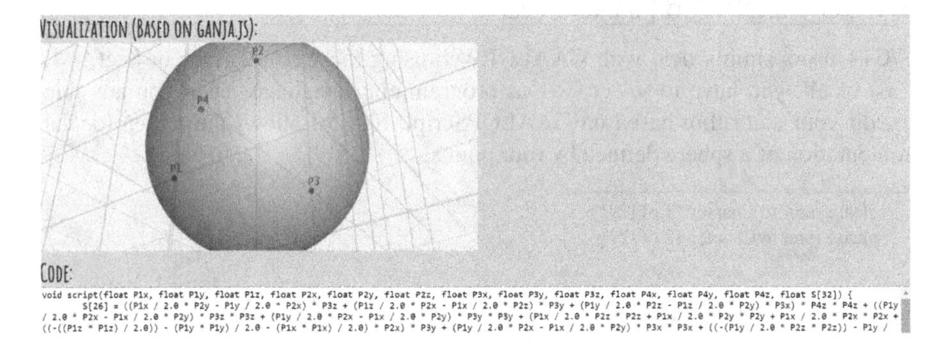

Figure 5.2 The result of the sphere-center example.

The code area presents the optimized C code as a C function with the parameters P1x, P1y, P1z, P2x, P2y, P2z, P3x, P3y, P3z, P4x, P4y and P4z. It computes the 3 coordinates of the center of the sphere as C[1] ..C[3]. They are returned by the float array C[32]. This code can simply be used in your C/C++ environment based on copy and paste. The visualization area shows the sphere S as well as the four points defining it. They are based on the fixed coordinates of the *Variable assignments* area.

```
Variable assignments:
1  P1x=0; P1y=0;
2  P2x=0; P2y=1; P2z=0;
3  P3x=1; P3y=0; P3z=0;
4  P4x=0; P4y=0.707; P4z=0.707;
5
```

Figure 5.3 The sphere-centre example without the P1z variable.

If variables should be possible to be adjusted dynamically, there is the option of using sliders. In this case the corresponding variables are omitted in the variable definition. For our example Fig. 5.3 shows the screen without the definition of the P1z variable. Fig. 5.4 shows the screen for the results.

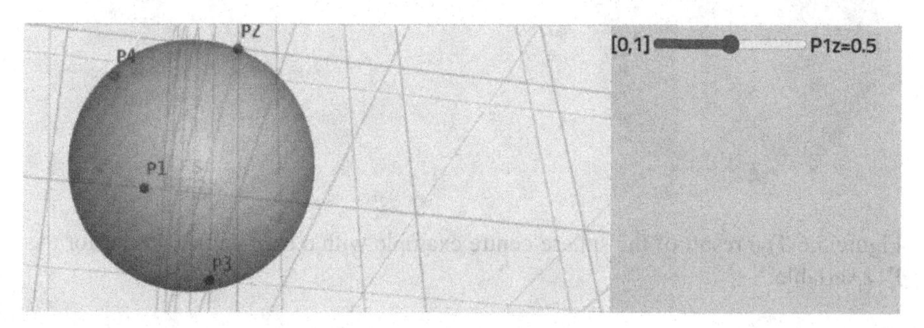

Figure 5.4 The result of the sphere-center example with a default slider for the P1z variable.

The visualization is now equipped with a slider for the P1z variable. The default range of a slider is [0..1]. If you would like to use another range, you are able to define it with a range pragma according to Fig. 5.5.

Variable assignments:

```
1  //#pragma range -0.3<=P1z<=0.3
2  P1x=0; P1y=0;
3  P2x=0; P2y=1; P2z=0;
4  P3x=1; P3y=0; P3z=0;
5  P4x=0; P4y=0.707; P4z=0.707;
6
```

Figure 5.5 The sphere-centre example with a user-defined slider definition.

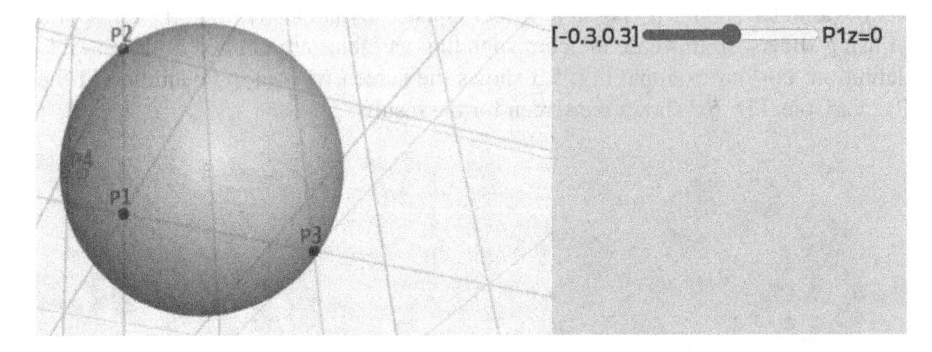

Figure 5.6 The result of the sphere-centre example with a user-defined slider for the P1z variable.

Now the slider is in the range of [-0.3 .. 0.3]. Please notice that the default value is always the middle of the range.

The GAALOPScript according to Listing 5.1 focuses on the efficient computation of the centre point of the sphere using the output pragma according to Table 3.4.

Listing 5.1: GAALOPScript for the computation of the centre of a sphere defined by 4 points.

```
1  P1 =  createPoint(P1x, P1y, P1z);
2  P2 =  createPoint(P2x, P2y, P2z);
3  P3 =  createPoint(P3x, P3y, P3z);
4  P4 =  createPoint(P4x, P4y, P4z);
5
6  S= P1^P2^P3^P4;
7
8  //#pragma output C e1 e2 e3
9  ?C = Normalize(S);
```

If a sphere is normalized[3], its e_1, e_2, e_3-components describe the centre point of the sphere [4], means the other coefficients do not have to be computed.

5.2 CODE GENERATION AND RUNTIME PERFORMANCE BASED ON GAALOPWeb

GAALOPWeb for C++ takes a Geometric Algebra algorithm in the form of a GAALOPScript and generates optimized C functions, which are optimized in terms of high runtime-performance. The GAALOPScript for the bisector example of Sect. 3.2.3 results in the following optimized C function:

```
void calculate(float x1,float x2,float y1,float y2,float L[16]){

L[1] = 2.0 * x2 - 2.0 * x1; // e1
L[2] = 2.0 * y2 - 2.0 * y1; // e2
L[3] = y2 * y2 - y1 * y1 + x2 * x2 - x1 * x1; // einf
}
```

We see that the coordinates x1, y1 and x2, y2 for the two points representing the line segment are taken and the line L representing the perpendicular bisector is computed. L is an array of the 16 coordinates as described in Table 2.6. Although six multivectors are assigned in the relevant GAALOPScript, only L is really computed in this function, since it is the only one indicated by a leading semicolon. Principally, GAALOPWeb is performing a symbolic precomputation of the coefficients of the multivectors and we can see that this is resulting in very simple elementary operations (additions and multiplications). To be a bit more precise, GAALOPWeb computes the multivectors symbolically and optimizes the calculations of each coefficient by symbolic simplifications. We stress that GAALOPWeb generates the computations for all non-zero coefficients of the multivector L.

The comments at the end of each assignment indicate that these are the coefficients of e_1, e_2 and e_∞. We also realize that only the variables x_1, x_2, y_1, y_2 for the coefficients of the 2D points are used. After the symbolic optimization of GAALOPWeb the variable r for the radius of the circles is no longer needed.

The optimization potential can be seen based on the following GAALOPScript, where all the statements have to be computed explicitly.

```
?P1 = createPoint(x1,y1);
?P2 = createPoint(x2,y2);

?S1 = P1 - 0.5*r*r*einf;
?S2 = P2 - 0.5*r*r*einf;

?PP = S1^S2;
```

[3]this is computed based on the predefined macro Normalize according to Table 3.3

[4]see, for instance, Sect. 5.6 of "Introduction to Geometric Algebra Computing" [30]

```
?L = *(*PP^einf);
```

Its result is the following C function:

```
void calculate(float r, float x1, float x2, float y1, float y2,
  float L[16], float P1[16], float P2[16],
  float PP[16], float S1[16], float S2[16]) {
P1[1] = x1; // e1
P1[2] = y1; // e2
P1[3] = (y1 * y1) / 2.0 + (x1 * x1) / 2.0; // einf
P1[4] = 1.0; // e0
P2[1] = x2; // e1
P2[2] = y2; // e2
P2[3] = (y2 * y2) / 2.0 + (x2 * x2) / 2.0; // einf
P2[4] = 1.0; // e0
S1[1] = P1[1]; // e1
S1[2] = P1[2]; // e2
S1[3] = P1[3] - (r * r) / 2.0; // einf
S1[4] = 1.0; // e0
S2[1] = P2[1]; // e1
S2[2] = P2[2]; // e2
S2[3] = P2[3] - (r * r) / 2.0; // einf
S2[4] = 1.0; // e0
PP[5] = S1[1] * S2[2] - S1[2] * S2[1]; // e1 ^ e2
PP[6] = S1[1] * S2[3] - S1[3] * S2[1]; // e1 ^ einf
PP[7] = S1[1] - S2[1]; // e1 ^ e0
PP[8] = S1[2] * S2[3] - S1[3] * S2[2]; // e2 ^ einf
PP[9] = S1[2] - S2[2]; // e2 ^ e0
PP[10] = S1[3] - S2[3]; // einf ^ e0
L[1] = (-PP[7]); // e1
L[2] = (-PP[9]); // e2
L[3] = (-PP[10]); // einf
}
```

Here, the needed coefficients of all multivectors have to be computed explicitly. There is also a need for computing with the variable *r*, since the optimization potential over the whole algorithm can not be used.

There is also another optimization possibility, if we use exclamation marks instead of the question marks.

```
!P1 = createPoint(x1,y1);
!P2 = createPoint(x2,y2);

!S1 = P1 - 0.5*r*r*einf;
```

```
!S2 = P2 - 0.5*r*r*einf;

!PP = S1^S2;

?L = *(*PP^einf);
```

In this case, coefficients are computed only if they are needed for further computations (see Table 3.1).

```
void script(float r, float x1, float x2, float y1, float y2,
  float L[16], float P1[16], float P2[16],
  float PP[16], float S1[16], float S2[16]) {
P1[1] = x1; // e1
P1[2] = y1; // e2
P1[3] = (y1 * y1) / 2.0 + (x1 * x1) / 2.0; // einf
P2[1] = x2; // e1
P2[2] = y2; // e2
P2[3] = (y2 * y2) / 2.0 + (x2 * x2) / 2.0; // einf
S1[1] = P1[1]; // e1
S1[2] = P1[2]; // e2
S1[3] = P1[3] - (r * r) / 2.0; // einf
S2[1] = P2[1]; // e1
S2[2] = P2[2]; // e2
S2[3] = P2[3] - (r * r) / 2.0; // einf
PP[7] = S1[1] - S2[1]; // e1 ^ e0
PP[9] = S1[2] - S2[2]; // e2 ^ e0
PP[10] = S1[3] - S2[3]; // einf ^ e0
L[1] = (-PP[7]); // e1
L[2] = (-PP[9]); // e2
L[3] = (-PP[10]); // einf
}
```

Now, only 18 assignments are generated compared to 25 assignments before.

This example is also used in order to illustrate a way to generate optimized Mathematica code according to Sect. 7.2.1.

CHAPTER 6

GAALOPWeb for Python

CONTENTS

There are two ways for Python users to deal with GAALOPWeb. One is simply using its web interface according to Sect. 6.1. For users who would like to have a solution really contained within the Python environment there is a second way according to Sect. 6.2. For Python, there are also the packages Clifford and Pyganja available according to Sect. 6.3. The integration of GAALOPWeb into a Python environment based on these packages is shown in Sect. 6.4.

6.1 THE WEB INTERFACE

In principle, the web interface of GAALOPWeb is presented in Sect. 4.1. First of all, you have to select Python as programming language. Then you are able to edit your algorithm based on GAALOPScript. Figure 6.1 shows the screen for the line-sphere example[1].

[1] see the corresponding GAALOPScript in Sect. 4.3.3

DOI: 10.1201/9781003139003-6

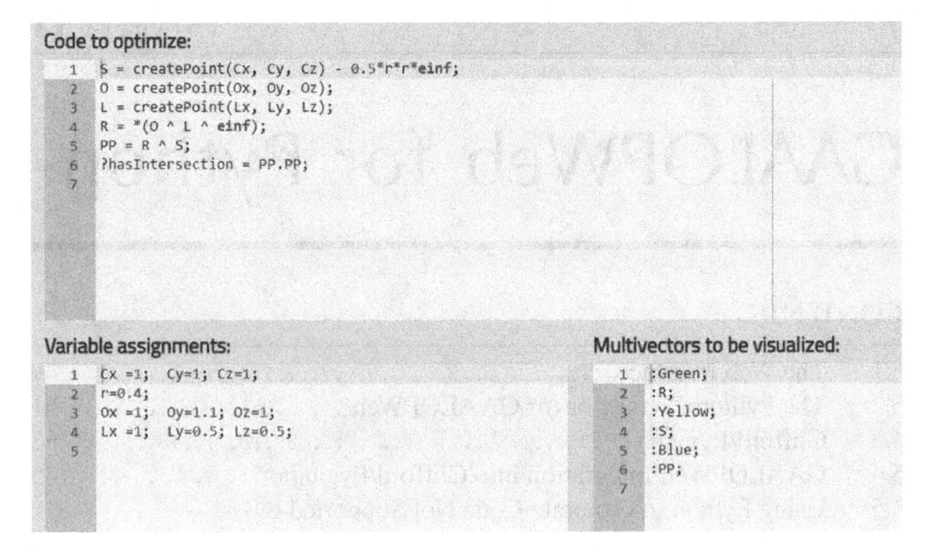

Figure 6.1 The line-sphere example with GAALOPWeb.

The general algorithm acccording to Sect. 3.2.4 can be seen in the *code to optimize area*. The visualization according to Sect. 4.3.3 is defined in the *Variable assignments* and the *Multivectors to be visualized* area. Fig. 6.2 shows the screen for the results of the line-sphere example.

The visualization area shows the sphere S together with the line *R* and the intersection point pair *PP*. The code area presents the optimized Python code as a Python function. This code can simply be used in your Python environment based on copy and paste. Listing 6.1 shows the Python code in some more detail.

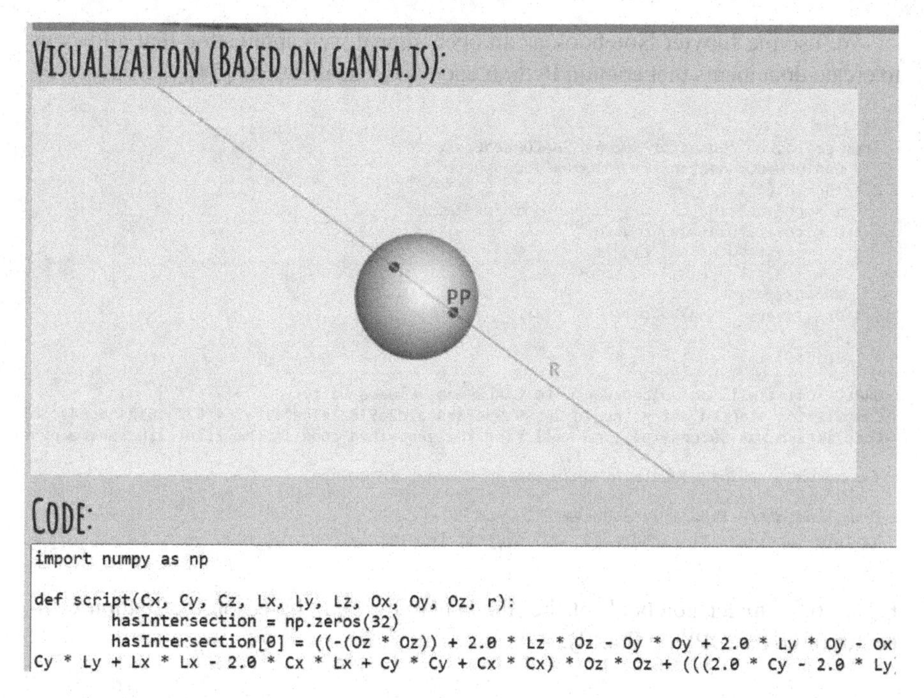

Figure 6.2 The result of the line-sphere example with GAALOPWeb.

Listing 6.1: The generated Python code of the line-sphere example.

```
1  import numpy as np
2
3  def script(Cx, Cy, Cz, Lx, Ly, Lz, Ox, Oy, Oz, r):
4          hasIntersection = np.zeros(32)
5          hasIntersection[0] = ((-(Oz * Oz)) ...   ; # 1.0
6          return hasIntersection
```

Since the multivectors are represented as numpy arrays, this package has to be imported. The generated function has the parameters Cx, Cy, Cz, Lx, Ly, Lz, Ox, Oy, Oz and r. The coefficients for the computed sphere are returned by the 32D array hasIntersection. From this array only the element with index 0 is used since only the scalar part of the multivector is needed. This is indicated by the comment at the end of the corresponding assignment. In this case this is only the scalar indicated by

`# 1.0`.

6.2 THE PYTHON CONNECTOR FOR GAALOPWeb

For users who would like to have a solution contained within the Python environment there is a second way for a more direct solution based on the Python Connector to be downloaded from the web page (see Fig. 4.2).

We use the Jupyter Notebook as an open-source web application that allows us to create documents that contain Python code and visualizations [54].

```
from gaalopweb_connector import GAALOPWebConnector
g = GAALOPWebConnector(scriptName="linesphere")
g.setScript('''
    S = createPoint(Cx, Cy, Cz) - 0.5*r*r*einf;
    O = createPoint(Ox, Oy, Oz);
    L = createPoint(Lx, Ly, Lz);
    R = *(O ^ L ^ einf);
    PP = R ^ S;
    ?hasIntersection = PP.PP;
''')
g.compile()
```

```
GAALOPScript will be uploaded now to GAALOPWeb! Please wait.
Compilation started using the id 26b89c99ef8edca2d2a5182244be1618e764067c6b61245ec64ac58!
Compilation was successful. You will find the generated code in the file: linesphere.py
```

You can easily call the generated function using the following commands:

```
from linesphere import linesphere
hasIntersection = linesphere(Cx, Cy, Cz, Lx, Ly, Lz, Ox, Oy, Oz, r)
```

Figure 6.3 Jupyter notebook of the line-sphere example based on the Python Connector of GAALOPWeb (Part I).

First, we have to make a connection for a GAALOPScript. In Fig. 6.3 the connection is called g and the script name is *linesphere*. Then, we have to set the text of the GAALOPScript and compile it to an optimized Python function[2]. We are automatically informed about how we have to call this function.

```
Cx=1;Cy=1;Cz=0;
r=0.4;
Ox=1;  Oy=1.1; Oz=1;
Lx=1;  Ly=0.5; Lz=0.5;

from linesphere import linesphere
hasIntersection = linesphere(Cx, Cy, Cz, Lx, Ly, Lz, Ox, Oy, Oz, r)
print (hasIntersection[0])
if hasIntersection[0] > 0:
    print("there is an intersection")
else:
    print("there is no intersection")
```

```
0.20490000000000008
there is an intersection
```

Figure 6.4 Jupyter notebook of the line-sphere example based on the Python Connector of GAALOPWeb (Part II).

[2]this part of the Python program has to be run only in the case of changes in the GAALOPScript

This integration is used in the program part of Fig. 6.4 together with the assignment of some concrete values and some print functionality. We can see that for these values there is an intersection since the value of *hasIntersection* is positive.

Figure 6.5 shows how visualizations can be integrated.

```
g.setVisualizationAssignments('''
Cx=1;Cy=1;Cz=1;
r=0.4;
Ox=1;  Oy=1.1; Oz=1;
Lx=1;  Ly=0.5; Lz=0.5;
''')
exec(g._visualizationAssignments)
g.setMvToBeVisualized(":Green;:S;:Blue;:R;:Red;:PP;")
g.visualize()
```

```
GAALOPScript will be uploaded now to GAALOPWeb! Please wait.
Compilation started using the id 9efae833f12672a01df2c3581aa6fc3b71e75f337769104269f87e1! Please wait for the result.
Compilation was successful. You will find the generated visualization below!
```

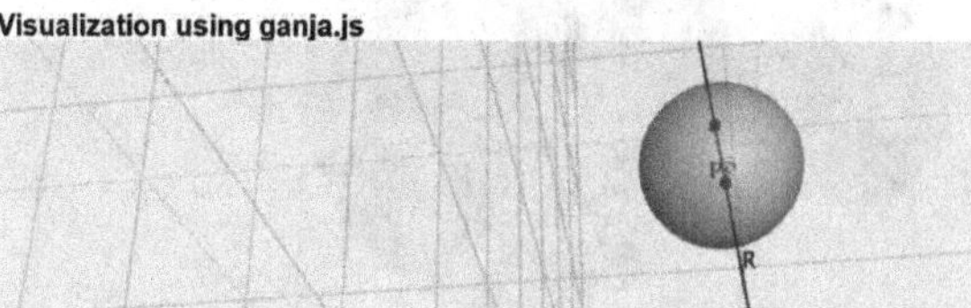

Figure 6.5 Jupyter notebook for the visualization of the line-sphere example.

6.3 CLIFFORD/PYGANJA

For Python, there are also the packages Clifford and Pyganja [5]. Clifford is a library for the computation with Geometric Algebra and Pyganja is a Python interface to ganja[3] in order to visualize with Geometric Algebra.

Listing 6.2 imports these packages, computes some geometric objects, reflects and visualizes them.

Listing 6.2: example Python program with Clifford and pyganja.

```
1  # Import Conformal GA (4,1)
2  from clifford.g3c import *
3  # Import prebuilt tools for conformal GA
4  from clifford.tools.g3c import *
5
6  from pyganja import *
7
8  point = up(2*e1+e2)
9  line = up(3*e1 + 2*e2) ^ up(3*e1 - 2*e2) ^ einf
10 circle = up(e1) ^ up(-e1 + 1.6*e2 + 1.2*e3)
11          ^ up(-e1 - 1.6*e2 - 1.2*e3)
12 sphere = up(3*e1) ^ up(e1) ^ up(2*e1 + e2) ^ up(2*e1 + e3)
13
14 # note that due to floating point rounding, we need to
15 # truncate back to a single grade here, with ``(grade)``
16 point_refl = homo((circle * point.gradeInvol() * ~circle)(1))
17 line_refl = (circle * line.gradeInvol() * ~circle)(3)
```

[3] see Sect. 2.4

```
18 sphere_refl = (circle * sphere.gradeInvol() * ~circle)(4)
19
20 draw([circle, line, sphere,point_refl, line_refl, sphere_refl]
21        , static=False)
```

The visualization result is shown in Fig. 6.6.

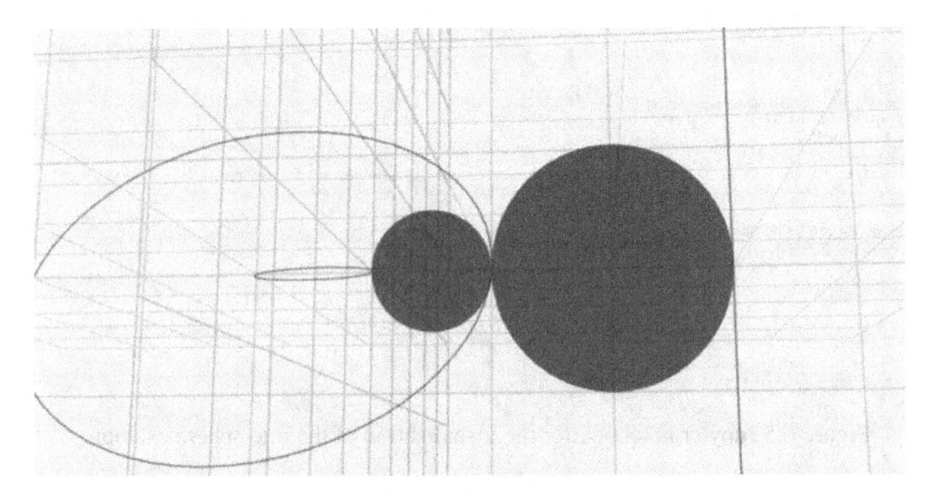

Figure 6.6 Simple visualization of some geometric objects with pyganja.

In order to visualize the objects with colour and labels, scene graphs have to be defined. Listing 6.3 defines a scene sc with the objects as well as a scene Sc_refl with the reflected ones.

Listing 6.3: Python program with Scenegraph extension.

```
1  sc = GanjaScene()
2  sc.add_object(point, color=(255, 0, 0), label='point')
3  sc.add_object(line, color=(0, 255, 0), label='line')
4  sc.add_object(circle, color=(0, 0, 255), label='circle')
5  sc.add_object(sphere, color=(0, 255, 255), label='sphere')
6
7  sc_refl = GanjaScene()
8  sc_refl.add_object(point_refl,
9       color=(128, 0, 0), label='point_refl')
10 sc_refl.add_object(line_refl.normal(),
11      color=(0, 128, 0), label='line_refl')
12 sc_refl.add_object(sphere_refl.normal(),
13      color=(0, 128, 128), label='sphere_refl')
14
15  draw(sc + sc_refl, scale=0.5)
```

The resulting visualization of Fig. 6.7 shows the geometric objects in different colours and labeled with their names.

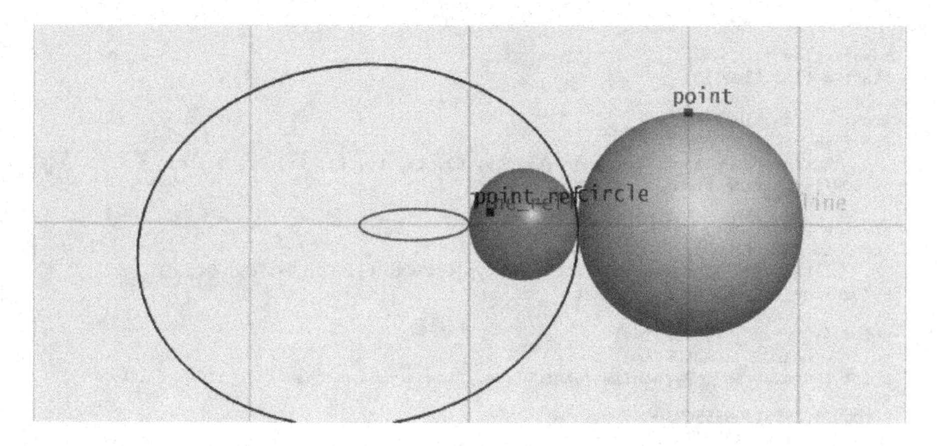

Figure 6.7 Pyganja scenegraph visualization.

6.4 GAALOPWeb INTEGRATION INTO CLIFFORD/PYGANJA

For a local GAALOP installation there is a solution for the automatic integration into the Python environment called gajit[25]. Using GAALOPWeb, the data structure of the generated multivector is compatible to the Clifford multivector. This is why GAALOPWeb can be seamlessly integrated into Clifford/Pyganja.

The line-sphere example of the previous sections can be expressed based on the Clifford library according to Listing 6.4.

Listing 6.4: The line-sphere example based on the Clifford library.

```
1  from clifford.g3c import *
2
3  def createPoint(x,y,z):
4      return up(x*e1 + y*e2 + z*e3)
5
6  def dual (mv):
7      return (e1^e2^e3^einf^eo)*mv
8
9  def linesphereClifford(Cx, Cy, Cz, Lx, Ly, Lz, Ox, Oy, Oz, r):
10     S = createPoint(Cx, Cy, Cz) - 0.5*r*r*einf
11     O = createPoint(Ox, Oy, Oz)
12     L = createPoint(Lx, Ly, Lz)
13     R = dual(O ^ L ^ einf)
14     PP = R ^ S
15     hasIntersection = PP|PP
16     return hasIntersection
```

The function *linesphereClifford()* is doing exactly the same as the function *linesphere()* generated by GAALOPWeb according to Fig. 6.4. In order to make the two descriptions (see GAALOPScript Listing 3.3) more compatible we introduce the functions *createPoint()* and *dual()*.

Now we are able to compare the two implementations according to the Python code of Fig. 6.8.

```
import time
start = time.time()

start = time.time()
for i in range(1000):
        hasIntersection = linesphere(Cx, Cy, Cz, Lx, Ly, Lz, Ox, Oy, Oz, r)
GAALOPTime = time.time() - start

start = time.time()
for i in range(1000):
        intersec = linesphereClifford(Cx, Cy, Cz, Lx, Ly, Lz, Ox, Oy, Oz, r)
PythonTime = time.time() - start

print ("Python", PythonTime)
print ("GAALOP", GAALOPTime)
print ("GAALOP is", PythonTime/GAALOPTime, "times faster than the Clifford library")
```

```
Python 0.2563130855560303
GAALOP 0.01097249984741211
GAALOP is 23.359588892268913 times faster than the Clifford library
```

Figure 6.8 Jupyter notebook of the line-sphere example based on the Python Connector of GAALOPWeb (Part II).

We can see that the code which is optimized by GAALOPWeb is about 20 times faster than the Clifford code.

6.5 USING PYTHON TO GENERATE CODE NOT SUPPORTED BY GAALOPWeb

In case of a programming language where no GAALOP/GAALOPWeb backend is available, there is also an alternative solution. As an example, we take some C code generated by GAALOPWeb and write a tool in order to translate it into Mathematica code[4]. Given that Python has become the language of choice for many engineering areas, we use the Python script according to Appendix A for that. The "Bisector" example of Sect. 3.2.3 shall be the basis for the description of our process for generating optimized code for Mathematica. First of all, GAALOPWeb generates the following optimized C function:

```
void calculate(float x1,float x2,float y1,float y2,float L[16]){

L[1] = 2.0 * x2 - 2.0 * x1; // e1
L[2] = 2.0 * y2 - 2.0 * y1; // e2
L[3] = y2 * y2 - y1 * y1 + x2 * x2 - x1 * x1; // einf
}
```

Our automatic conversion to Mathematica based on the Python script of Appendix A produces:

[4]this is just an example, since optimized code for Mathematica can already be generated according to Sect. 7.2.1

```
Bisector[x1_, x2_, y1_, y2_] := Module[{L},
L = ConstantArray[0, 16];
L[[2]] = 2.0*x2 - 2.0*x1 (*e1*);
L[[3]] = 2.0*y2 - 2.0*y1 (*e2*);
L[[4]] = y2*y2 - y1*y1 + x2*x2 - x1*x1 (*einf*);
Return[L];]
```

The Python script is taking the optimized C file and translating it to a file with correct Mathematica syntax

- take the name of the GAALOPScript for the Mathematica function

- define a constantArray for the multivector to be computed

- handle the array indices correctly for Mathematica (when GAALOP generates C/C++ code, the index for e_1 is 1, because the indexes start from 0, with the scalars. Mathematica doesn't have the 0 index, so the scalar representative index is 1.)

- use (* and *) for the comments indicating which blade is corresponding to the relevant array entry

etc.

The function created in Mathematica has a "Module" structure, which protects the variable names. In cr4d all the elements are arrays with 16 coordinates; in 5d conformal model, for instance, the elements are arrays with 32 coordinates. All of them are initiated as arrays of zeros.

Appendix A shows the Python script for the conversion of optimized C code to Mathematica code. The function replaceFunctionCall replaces C functions such as sqrtf with Mathematica functions such as Sqrt and handles the specific brackets accordingly. GAALOPtoMathematica is the main function going line per line through the optimized C program and converting it to the correct Mathematica syntax.

CHAPTER 7

Molecular Distance Application Using GAALOPWeb for Mathematica

CONTENTS

GAALOPWeb for Mathematica supports the Mathematica user with an intuitive interface for the development, testing and visualization of Geometric Algebra algorithms, combining the geometric intuitiveness of Geometric Algebra with an efficient development of algorithms for Wolfram Mathematica. We particularly illustrate this integration using an implementation of a Distance Geometry Problem, which consists of finding three-dimensional embeddings of graphs. We mainly follow [2].

Mathematica is a software developed by Stephen Wolfram around 1980, and is a widely used and very powerful tool. It has many implemented functions, a great documentation, and it is suitable for symbolic computation [67]. There are already Geometric Algebra implementations for Mathematica, the GrassmannAlgebra package [7], the CGA (Conformal Geometric Algebra)-Package of Kondo et al [45], which is a package for 5D CGA used to solve an origami problem; the Geometric Algebra library of Terje Vold, a package for three-dimensional space and four-dimensional spacetime computations [65]. Mathematica is also used to perform symbolic computations for GMac, a tool for generating optimized Geometric Algebra code for C# among other languages, developed by Ahmad Eid [1].

DOI: 10.1201/9781003139003-7

We use a Distance Geometry application (see Sect. 7.1) for illustration purposes of this chapter. We focus in Section 7.2 on GAALOPWeb and its efficient development of Geometric Algebra algorithms for Mathematica. In Sect 7.3 we use our Distance Geometry application in order to compare a conventional solution with our approach in terms of runtime performance and accuracy.

7.1 DISTANCE GEOMETRY EXAMPLE

The application chosen for illustration purposes is the Molecular Distance Geometry Problem (MDGP) [52], which is one of the main problems in Distance Geometry. In recent years, the MDGP has been addressed through Geometric Algebra, especially through the conformal model, in a very intuitive way, and the advances can be seen in [3, 9, 17, 47, 48]. It consists of finding a three-dimensional structure of a molecule given an incomplete set of interatomic distances. This is classically treated as a global optimization problem, but it can be discretized if it is possible to find an ordering in the sequence of atoms satisfying some conditions. This class of problems is called the Discretizable MDGP (DMDGP) and it is defined below.

Definition of DMDGP Given a simple undirected graph $G=(V,E,d)$, where V is the set of vertices, E the set of edges and $d: E \to (0,\infty)$ a distance function on V; and an order on the vertices V, denoted by $v_1,...,v_n$, such that:

1. $\{v_1,v_2,v_3\}$ are fixed and each vertex v_i, $i>3$, is adjacent to at least three consecutive predecessors, i.e.

$$\forall i>3, \{v_{i-3},v_i\}, \{v_{i-2},v_i\}, \{v_{i-1},v_i\} \in E,$$

2. the distances among v_{i-3},v_{i-2},v_{i-1} satisfy the triangle inequality

$$d_{i-3,i-2}+d_{i-2,i-1} > d_{i-3,i-1},$$

find an embedding $x: V \to \mathbb{R}^3$, such that

$$\forall \{v_i,v_j\} \in E, \quad \|x_i - x_j\| = d_{i,j}. \tag{7.1}$$

The DMDGP describes the protein backbone. Some of the distances come from the protein structure itself, as bond lengths and bond angles, which are considered as exact real numbers, and some data are obtained from experiments of Nuclear Magnetic Resonance (NMR) [49]. In general, the experimental data is not exact and is considered as an interval, called interval distances.

A *branch-and-prune* (BP) algorithm can be used to solve the DMDGP [10, 51, 56]. The idea of BP is to search possible positions at each level (branching) and test its feasibility (pruning) when distances $d_{i,j}$, $j-i>3$, are available. Geometrically, the search for an atom x_i can be interpreted as an intersection of three spheres centred in the predecessors x_{i-1}, x_{i-2} and x_{i-3} with radius $d_{i-1,i}$, $d_{i-2,i}$ and $d_{i-3,i}$, respectively. This is the main motivation for using CGA. Classically, the branching phase of BP uses 4×4 matrices multiplications to compute sphere intersections

[46]. Although not intuitive, this technique avoids solving the nonlinear system of equations associated to (7.1). Another possibility is to linearize (7.1) at each step of branching by subtracting one equation from the others. The linearization works like this: consider the system of nonlinear equations below, where the points x_{i-3}, x_{i-2} and x_{i-1} are fixed and we want to find the point x_i. The distances $d_{i-3,i}$, $d_{i-2,i}$ and $d_{i-1,i}$ are given.

$$\begin{aligned} ||x_i - x_{i-3}||^2 &= d_{i-3,i}^2 \\ ||x_i - x_{i-2}||^2 &= d_{i-2,i}^2 \\ ||x_i - x_{i-1}||^2 &= d_{i-1,i}^2 \end{aligned} \tag{7.2}$$

Subtracting, for instance, the first equation from the others, it becomes

$$\begin{aligned} ||x_i - x_{i-2}||^2 - ||x_i - x_{i-3}||^2 &= d_{i-2,i}^2 - d_{i-3,i}^2 \\ ||x_i - x_{i-1}||^2 - ||x_i - x_{i-3}||^2 &= d_{i-1,i}^2 - d_{i-3,i}^2, \end{aligned}$$

and this system can be written as a linear system in x_i. Indeed, we have

$$\begin{aligned} ||x_i||^2 + ||x_{i-2}||^2 - 2\langle x_i, x_{i-2}\rangle - ||x_i||^2 - ||x_{i-3}||^2 + 2\langle x_i, x_{i-3}\rangle &= d_{i-2,i}^2 - d_{i-3,i}^2 \\ ||x_i||^2 + ||x_{i-1}||^2 - 2\langle x_i, x_{i-1}\rangle - ||x_i||^2 - ||x_{i-3}||^2 + 2\langle x_i, x_{i-3}\rangle &= d_{i-1,i}^2 - d_{i-3,i}^2, \end{aligned}$$

where $\langle\cdot,\cdot\rangle$ is the scalar product in $\mathbb{R}^3$, and then

$$\begin{aligned} 2\langle x_i, x_{i-3} - x_{i-2}\rangle &= ||x_{i-3}||^2 - ||x_{i-2}||^2 + d_{i-2,i}^2 - d_{i-3,i}^2 \\ 2\langle x_i, x_{i-3} - x_{i-1}\rangle &= ||x_{i-3}||^2 - ||x_{i-1}||^2 + d_{i-1,i}^2 - d_{i-3,i}^2, \end{aligned}$$

which is the system of linear equations $Ax_i = b$, where $A = \begin{pmatrix} x_{i-3} - x_{i-2} \\ x_{i-3} - x_{i-1} \end{pmatrix}$, and $b = \frac{1}{2}\begin{pmatrix} ||x_{i-3}||^2 - ||x_{i-2}||^2 + d_{i-2,i}^2 - d_{i-3,i}^2 \\ ||x_{i-3}||^2 - ||x_{i-1}||^2 + d_{i-1,i}^2 - d_{i-3,i}^2 \end{pmatrix}$.

A hypothesis of the DMDGP ordering is that three consecutive atoms are not collinear, so the matrix A has full rank, and we can find the solutions for $Ax_i = b$. As this is a 2×3 system, the set of solutions depends on a parameter t.

To find the possible values for t we solve an equation of the original system (7.2) for $x_i(t)$, like

$$||x_i(t) - x_{i-3}||^2 = d_{i-3,i}^2, \tag{7.3}$$

to get two solutions t_1 and t_2. The possible positions for the atom x_i are given by $x_i(t_1)$ and $x_i(t_2)$.

If the distance $d_{i,i-3}$ is exact, the intersection (if it exists) is a point pair. However, if $d_{i,i-3}$ is an interval, the possible positions are in a circular arc in the intersection of two spheres and a spherical shell [3, 10]. An important issue of BP is that for uncertain data, it discretizes the intervals to continue the search. It possibly leads to an unfeasible problem, because only a few points in the interval are considered and they can be all unfeasible; or to a huge increase in the dimension, once many points are considered at each interval and each one is used to the next level.

As we mentioned in Sect. 2.2, in CGA we can compute with spheres and perform intersections between them very intuitively compared to the linearization depicted above. Again, let x_{i-3}, x_{i-2} and x_{i-1} be fixed points and $d_{i-3,i}$, $d_{i-2,i}$ and $d_{i-1,i}$ their given distances to the atom x_i, we are able to compute the three sphere intersections as follows:

$$S_{i-3} = X_{i-3} - \frac{1}{2}d_{i-3,i}^2 e_\infty$$
$$S_{i-2} = X_{i-2} - \frac{1}{2}d_{i-2,i}^2 e_\infty$$
$$S_{i-1} = X_{i-1} - \frac{1}{2}d_{i-1,i}^2 e_\infty,$$

where $X_{i-j} = x_{i-j} + \frac{1}{2}x_{i-j}^2 e_\infty + e_0$. Then, we use the outer product to get the intersection

$$Pp_i = S_{i-3} \wedge S_{i-2} \wedge S_{i-1},$$

and finally extract the points from the point-pair

$$X_i(-) = \frac{-\sqrt{(Pp_i^*)^2} + Pp_i^*}{-e_\infty \cdot Pp_i^*}, \qquad X_i(+) = \frac{\sqrt{(Pp_i^*)^2} + Pp_i^*}{-e_\infty \cdot Pp_i^*}.$$

We will see in Sect. 7.2 how easy it is to implement this algorithm efficiently in Mathematica based on GAALOPWeb.

7.2 GAALOPWEB FOR MATHEMATICA

GAALOPWeb for Mathematica is an easy to handle online solution (without any software installation needed) including a generator for Mathematica code as well as an integrated visualization.

7.2.1 Mathematica Code Generation

GAALOPWeb generates the following Mathematica code based on the "Bisector" example in Sect. 3.2.3:

```
Bisector[x1_, x2_, y1_, y2_] := Module[{L},
L = ConstantArray[0, 16];
L[[2]] = 2.0*x2 - 2.0*x1 (*e1*);
L[[3]] = 2.0*y2 - 2.0*y1 (*e2*);
L[[4]] = y2*y2 - y1*y1 + x2*x2 - x1*x1 (*einf*);
Return[L];]
```

The code

- has a "Module" structure, which protects the variable names;

- defines a constantArray for the multivector L to be computed (in cr4d all the elements are arrays with 16 coordinates; in the 5D conformal model, for instance, the elements are arrays with 32 coordinates. All of them are initiated as arrays of zeros);

- handles the array indices correctly for Mathematica (when GAALOP generates C/C++ code, the index for e_1 is 1, because the indices start from 0, with the scalars. Mathematica does not have the 0 index, so the scalar representative index is 1);

- uses (* and *) for the comments indicating which blade is corresponding to the relevant array entry; and

- handles the specific brackets for Mathematica.

In order to illustrate the GAALOPWeb-Mathematica integration we use the three sphere intersection example in Sect. 7.1. We define the conformal centres and their corresponding spheres, compute the point-pair and extract its points. The question marks in the script below indicate the elements to appear explicitly in the Mathematica code.

```
//creating the CGA points;
x1=createPoint(a1,a2,a3);
x2=createPoint(b1,b2,b3);
x3=createPoint(c1,c2,c3);

// creating the spheres;
?S1=x1-0.5*(d14*d14)*einf;
?S2=x2-0.5*(d24*d24)*einf;
?S3=x3-0.5*(d34*d34)*einf;

// The PointPair in the intersection;
?PP4=S1^S2^S3;
?DualPP4=*PP4;

// Extraction of the two points;
?x4a=-(-sqrt(DualPP4.DualPP4)+DualPP4)/(einf.DualPP4);
?x4b=-(sqrt(DualPP4.DualPP4)+DualPP4)/(einf.DualPP4);
```

Using the above GAALOPScript, we generate a Mathematica function to intersect three spheres and return the two points. Everything is optimized symbolically in GAALOPWeb and a single function is generated for Mathematica. We recall that this function returns 32-coordinate multivectors, but in this case it only computes 4: e_1, e_2, e_3 and e_∞ since the coefficient for e_0 is equal to 1 and all the others are 0. The generated Mathematica code starts as follows:

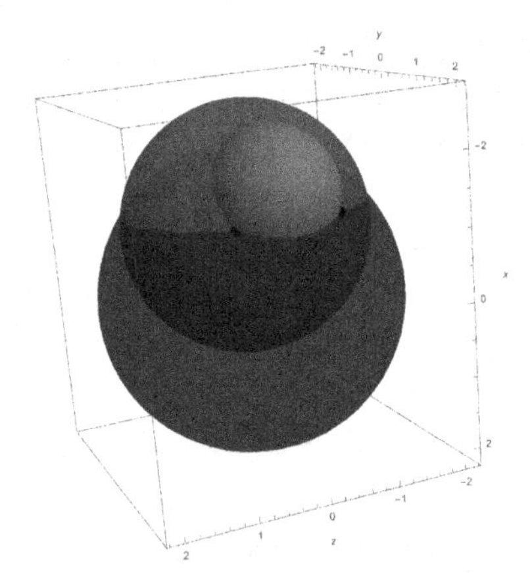

Figure 7.1 Figure generated in Mathematica for the example above. The points in black are exactly the points found by the CGA function.

```
CGAThreeSphereIntersection[{a1_, a2_, a3_},
{b1_, b2_, b3_}, {c1_, c2_, c3_}, d14_, d24_, d34_] :=Module
[{DualPP4, PP4, S1, S2, S3, x4a, x4b},...
```

For the points $x_1 = (0,0,0)$, $x_2 = (-1,0,0)$, and $x_3 = (-1.5, \frac{\sqrt{3}}{2}, 0)$, and distances $d_{1,4} = 2.15$, $d_{2,4} = \sqrt{3}$ and $d_{3,4} = 1$, an example of output is given e. g. by

```
{x4a,x4b}=CGAThreeSphereIntersection[x1,x2,x3,d14,d24,d34]
```

```
{{0,-1.31125,1.55235,-0.702375,2.31125,1.,0,0,0,0,0,0,0,0,0,0,0.,0.,0,0.,0,0,0.,0,0,0,0,0,0,
0,0,0,0,0},
{0,-1.31125,1.55235,0.702375,2.31125,1.,0,0,0,0,0,0,0,0,0,0,0.,0.,0,0.,0,0,0.,0,0,0,0,0,0,0,
0,0,0,0}}
```

In order to implement this function inside BP, we are interested in the coordinates $\{2,3,4\}$ of each point, which correspond to the coordinates of the points in $\mathbb{R}^3$.

The solutions can easily be checked with the graphics tools in Mathematica. The macros "Sphere" and "Point" as input of another macro "Graphics3D", draw the objects. The solution for the example above is shown in Fig. 7.1.

7.2.2 The Web-Interface

GAALOPWeb for Mathematica is simply a web page[1] for the definition and visualization of Geometric Algebra algorithms as well as to generate optimized code for Mathematica. In this paper, we will use CGA with GAALOPWeb. Please note, that GAALOPWeb is able to support various Geometric Algebras, not only the CGA. As mentioned before, a benefit for the usage of GAALOPWeb is, that the user has not to install GAALOP or Maxima, but just opens the website and focuses on the development of the GAALOPScript algorithm. This can also be done via a smart-phone or tablet.

[1] *www.gaalop.de/gaalopweb/res/mathematica/*

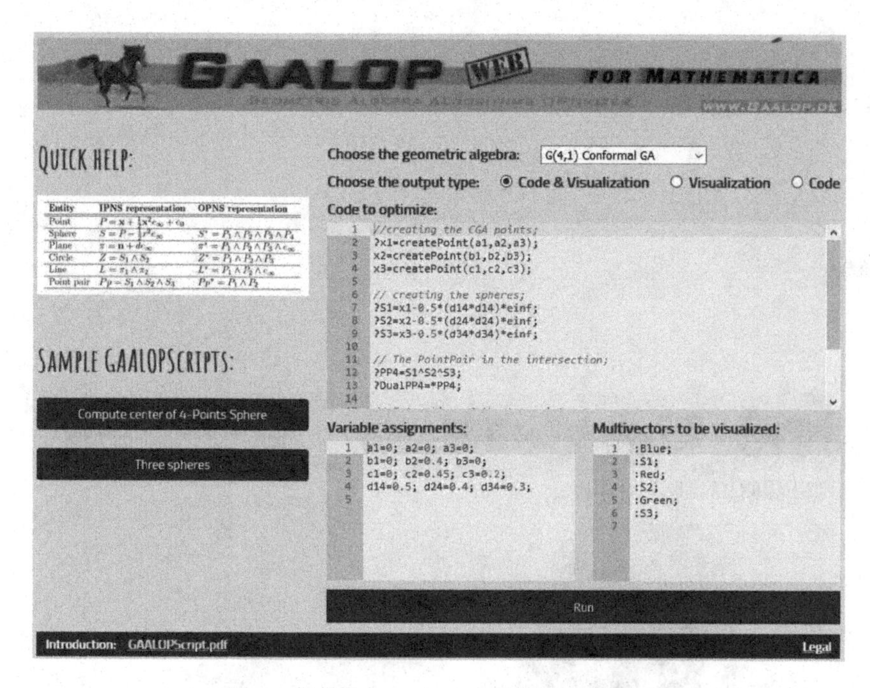

Figure 7.2 Screenshot of GAALOPWeb.

Figure 7.2 shows a screenshot with the three sphere intersection example of Sect. 7.2.1. On the top, a geometric algebra can be chosen, in this case, the CGA. Then we can decide, what GAALOPWeb should generate: The code, the visualization, or both. Here we want to generate the code and the visualization. Then we input the GAALOPScript, which we want to optimize. For the purpose of visualization, which is based on the Ganja.js package (see Sect. 2.4), some visualization data has to be defined. There are two text areas

Variable assignments

Multivectors to be visualized

For that, Figure 7.2 shows also the assignment of concrete values for the variables according to the example of Sect. 7.2.1 in the lower area. Accordingly, sphere S1 is defined to be visualized in blue colour, sphere S2 in red colour, etc.

Pushing the "Run" button leads to the computation of optimized code for Mathematica, as well as the generation of the visualization. The result is given in Fig. 7.3 which shows the result including the visualization and the generated optimized Mathematica code.

7.3 COMPUTATIONAL RESULTS

To test the results for the Three Sphere Intersection function we used the BP (see Sect. 7.1) for exact distances implemented in the MDGP package, developed by Michael Souza [61]. The instances are generated by an artificial structure randomly

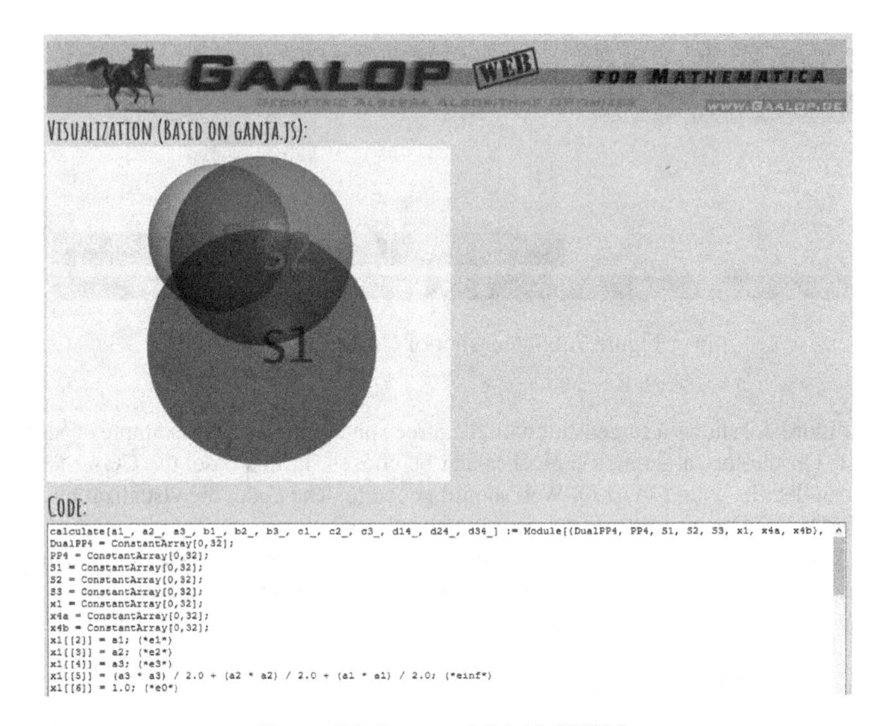

Figure 7.3 Output of GAALOPWeb.

created and by a function that retrieves only the distances $d_{i,i-1}$, $d_{i,i-2}$ and $d_{i,i-3}$. This function also maintains some additional distances less than a limit given by the user, the default value is 5^2, building a graph that satisfies the ordering of a MDGP instance. So the data available for the BP algorithm are the edges of this graph and their weight given by the distances.

In this package, the branching is done by linearizing the nonlinear system associated to the intersection, as described in Section 7.1. To find the three coordinates, the normal vector $\mathbf{n}$ of the plane generated by x_{i-3}, x_{i-2} and x_{i-1} is added to the matrix A, and the scalar product of $\mathbf{n}$ with x_{i-3} (or with some linear combination of x_{i-3}, x_{i-2} and x_{i-1}) is added to the vector b. Then we have,

$$A' = \begin{pmatrix} x_{i-3} - x_{i-2} \\ x_{i-3} - x_{i-1} \\ \mathbf{n} \end{pmatrix}$$

and

$$b' = \frac{1}{2} \begin{pmatrix} ||x_{i-3}||^2 - ||x_{i-2}||^2 + d^2_{i-2,i} - d^2_{i-3,i} \\ ||x_{i-3}||^2 - ||x_{i-1}||^2 + d^2_{i-1,i} - d^2_{i-3,i} \\ \langle \mathbf{n}, x_{i-3} \rangle, \end{pmatrix}$$

where $\mathbf{n} = \dfrac{(x_{i-3} - x_{i-2}) \times (x_{i-1} - x_{i-2})}{||(x_{i-3} - x_{i-2}) \times (x_{i-1} - x_{i-2})||}$, and $\times$ is the vector product.

The solution of the linear system $A'y = b'$ is on the plane and is the orthogonal projection of the intersection points. To find these points, it's necessary to compute the distance between them and y, and finally translate the projection through $\mathbf{n}$ to each side of the plane. The algorithm is described as follows.

[2] Angstrom is a unit of length

Data: ;
Points $x_{i-3}, x_{i-2}, x_{i-1}$;
Distances $d_{i-3,i}, d_{i-2,i}, d_{i-1,i}$
Result: Intersection points $x_i(a), x_i(b)$
begin

compute the $\mathbf{n}$, A and b;

$\mathbf{n} = (x_{i-3} - x_{i-2}) \times (x_{i-1} - x_{i-2})$;

$$A = \begin{pmatrix} x_{i-3} - x_{i-2} \\ x_{i-3} - x_{i-1} \\ \mathbf{n} \end{pmatrix}$$

$$b = \frac{1}{2} \begin{pmatrix} ||x_{i-3}||^2 - ||x_{i-2}||^2 + d_{i-2,i}^2 - d_{i-3,i}^2 \\ ||x_{i-3}||^2 - ||x_{i-1}||^2 + d_{i-1,i}^2 - d_{i-3,i}^2 \\ \langle \mathbf{n}, x_{i-3} \rangle \end{pmatrix}.;$$

$y = LinearSolve(A, b)$;

Compute the distance d_{y,x_i};

$d_{y,x_i} = ||y - x_i|| = \sqrt{||y - x_{i-3}||^2 - d_{i-3,i}^2}$;

return ;

$x_i(a) = y + d_{y,x_i}\mathbf{n}$;

$x_i(b) = y - d_{y,x_i}\mathbf{n}$;

end

Algorithm 1: Algorithm for the linear algebra approach for the intersection of three spheres.

In order to run our tests based on Geometric Algebra alternative, the function that implements Algorithm 1 is replaced by the function generated automatically from the GAALOPScript for the intersection of three spheres. For further optimization the script according to Section 7.2.1 is extended by two lines.

```
// Extraction of the two points;
//#pragma output x4a e1 e2 e3
//#pragma output x4b e1 e2 e3
?x4a=-(-sqrt(DualPP4.DualPP4)+DualPP4)/(einf.DualPP4);
?x4b=-(sqrt(DualPP4.DualPP4)+DualPP4)/(einf.DualPP4);
```

GAALOPWeb offers the possibility of restricting the coefficients of a multivector to be computed explicitly. Since we are only interested in the three Euclidean coordinates of the points $x4a$ and $x4b$, we use the output pragma in order to define that only the e_1, e_2, e_3-components of these points should be computed.

To perform intersections, it is important that the points are not collinear. To avoid situations where points are nearly collinear, the Algorithm 1 was implemented in such a way that the sequence of the centers can be changed in order to provide the most perpendicular angle.

Both functions were tested for the same 50 randomly generated instances, for 50, 80, 100 and 150 atoms. The resulting structures were compared to the original instance and the relative mean error (RME) between them was computed. Table 3 shows the means of the RME and of the running time for each case. Due to rounding errors it is possible that both methods find no intersection points at some step and consequently no solutions for the example.

TABLE 7.1 The Results for the Function *DGPSolver* [61] Running the Conventional and the CGA approach.

	# atoms	Found solutions	running time (s)	RME	RME both
BP conv	50	48	0.2022	6.4556×10^{-10}	=
BP GAALOP	50	48	0.2023	1.3984×10^{-9}	=
BP conv	80	35	0.3617	5.2523×10^{-9}	=
BP GAALOP	80	36	0.3617	4.2921×10^{-9}	4.0132×10^{-9}
BP conv	100	23	0.4840	4.9566×10^{-10}	=
BP GAALOP	100	25	0,4795	5.3293×10^{-9}	1.0821×10^{-9}
BP conv	150	11	0.7925	2.7701×10^{-9}	=
BP GAALOP	150	13	0.8185	1.8368×10^{-8}	5.734×10^{-9}

Each method was tested 50 times for examples randomly generated. The columns correspond respectively to: the number of atoms; the number of examples for which each method found solutions; mean running time for the examples when solutions were found; RME for the examples with solutions found for each method; RME for the examples when both methods found solutions. The sign "=" indicates equality to the column on the left.

We see that our approach provides competitive results in terms of runtime and accuracy compared to the conventional approach, and is capable to find solutions when the BP conventional did not. Concluding, this chapter shows for a distance geometry application that

- the description of algorithms in Geometric Algebra is easy and intuitive in Geometric Algebra compared to conventional algorithms;

- Mathematica implementations based on GAALOPWeb are efficient, based on a simple web-based interface supporting development and test with an easy description of algorithms and an integrated visualization; and

- the runtime performance and accuracy of the implementation is competitive compared to a conventional solution.

CHAPTER 8

Robot Kinematics Based on GAALOPWeb for MATLAB®

CONTENTS

In this chapter, we present *GAALOPWeb for MATLAB*, a new and easy to handle solution for Geometric Algebra implementations for MATLAB®. We demonstrate its usability for industrial applications based on a forward kinematics algorithm of a serial robot arm and illustrate it with the help of high run-time performance. We mainly follow [37].

We focus on the investigation of Conformal Geometric Algebra (CGA)[1] computations based on MATLAB which is very popular in the engineering and research community. For this purpose we originally used the Clifford Multivector Toolbox for MATLAB [59] in order to develop a kinematics algorithm of a real robot[2]. This toolbox provides a very intuitive way how MATLAB users can handle computations in Geometric Algebra. But, we realized that the runtime performance was far from being acceptable for industrial applications. This is why we developed GAALOPWeb for MATLAB as a tool for MATLAB users which is

easy to access

easy to use,

[1] see Sect. 2.2

[2] there is another tool for MATLAB called GABLE [53], but this library is no longer supported and its purpose is mainly educational

DOI: 10.1201/9781003139003-8

providing a high runtime performance for industrial applications.

Section 2.2 shows the high expressiveness of CGA as a mathematical language. As a sample manipulator we use the robot ABB IRB4400 and for the verification of the algorithm functionality we use ABB RobotStudio. The manipulator model is described in Section 8.1. Its kinematics[3] is demonstrated in Sect. 8.2 based on the forward kinematics problem. The problem itself is very suitable for Geometric Algebra. Please refer to [4] and [23] for not only forward but also inverse and differential kinematics. Yet the problem of a CGA implementation to a relevant engineering software remains. As mentioned above, we primarily used the Clifford Multivector Toolbox for MATLAB. This implementation is described in Sect. 8.3. The philosophy behind GAALOP (see Chapt. 3) is completely different to this solution. While a library-based solution is focusing on a good implementation of all library functions, GAALOP is taking the Geometric Algebra algorithm as a whole and generates an optimized implementation in terms of high runtime performance.

In order to achieve the above mentioned goals we extended GAALOP in a way to make it as easy as possible to handle for MATLAB users. The new way of generating optimized MATLAB code with the help of the web-based *GAALOPWeb for MATLAB* without the need of installing a specific software is presented in Sect. 8.5. The comparison of run-time performance according to Sect. 8.6 shows the power of this approach.

8.1 THE MANIPULATOR MODEL

Generally, a manipulator or a serial robot arm is a robotic device composed of various number of links. For our particular setting, links are denoted as l_{01}, l_{12}, l_{23} and l_{34}, see Figure 8.1. These links connect motorized joints $j_0, ..., j_3$ and the endpoint (gripper) j_4 which does not affect the manipulator kinematics. Also note that we consider only revolute joints whose setting is given by the values of appropriate angles $\varphi_1, ..., \varphi_5$ with the rotation direction indicated in Figure 8.1. Furthermore, the construction of the ABB IRB4400 manipulator in question connects the angle φ_2 with the angle between the links l_{12} and l_{23} in the sense that a modification of φ_2 causes an adjustment of the angle between l_{12} and l_{23} such that the orientation of the link l_{23} is preserved. Therefore, the rotation in φ_2 acts as a translation to the effector point. Figure 8.1 displays the initial position in which the value of all angles is set to 0.

8.2 KINEMATICS OF A SERIAL ROBOT ARM

A problem of forward kinematics consists of finding the final position of the endpoint j_4 based on given values of joint angles $\varphi_1, \dots, \varphi_5$. Our aim is to describe a kinematic chain that leads to the solution. We represent the joints in initial positions j_i, $i = 0, \dots, 4$ as CGA points $P_{j_i} = c(j_i)$, $i = 0, \dots, 4$, where $c()$ is the conformal

[3]for a short introduction to CGA-based serial robot arm kinematics we refer to [42]

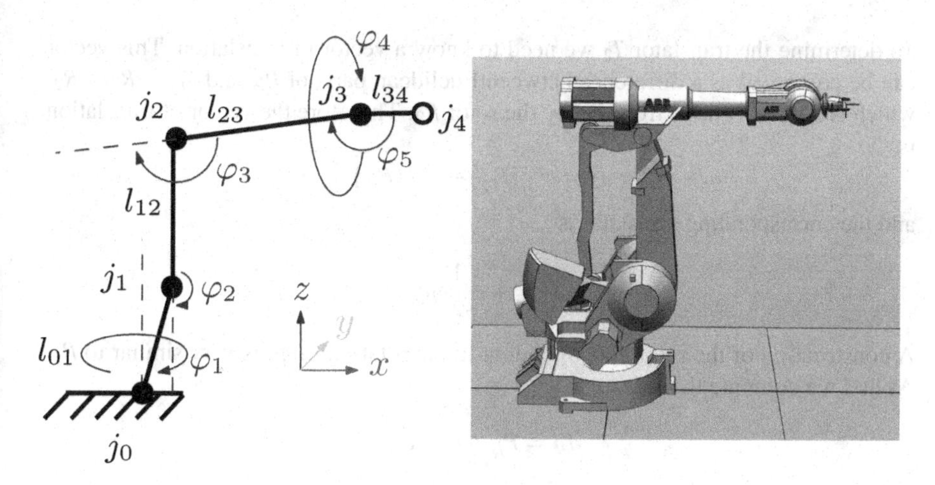

Figure 8.1 Manipulator model description.

embedding[4]. The final kinematic chain can be described in CGA as a composition of transformations.

To express the rotor[5] of the first transformation which is caused by φ_1, we need to construct the axis of rotation. Let $P_1 = c(j_0 + e_3)$ be a point shifted from j_0 by 1 in the z-axis direction. Then we define a line representing the z-axis (it is also the axis of rotation) as

$$L_1 = P_{j_0} \wedge P_1 \wedge e_\infty.$$

Now we can express the rotor corresponding to φ_1 as

$$R_1 = \cos\frac{\varphi_1}{2} + L_1^* \sin\frac{\varphi_1}{2},$$

where the star symbol stands for the duality since we require the axis of rotation to be in IPNS representation. The plus sign in the previous equation denotes that the direction of the rotation is clockwise.

A change of the angle φ_2 acts on the joint P_{j_2} as a rotation, however on P_{j_4} it acts as a translation since the orientation of the link l_{23} is preserved unless φ_3 or φ_1 is changed. As a consequence links l_{23} and l_{34} are translated. Therefore, we have to compute the rotor R_2 and the translator T_2. The axis of rotation is perpendicular to a robot's plane and it passes through P_{j_1}, i.e.

$$L_2 = P_{j_1} \wedge P_2 \wedge e_\infty,$$

where $P_2 = c(j_1 + e_2)$ is a point shifted from j_1 by 1 in the y-axis direction. The rotor corresponding to φ_2 is

$$R_2 = \cos\frac{\varphi_2}{2} + L_2^* \sin\frac{\varphi_2}{2}.$$

[4]A Euclidean vector $x \in \mathbb{R}^3$ in the embedding will be expressed as $x = x_1e_1 + x_2e_2 + x_3e_3$ so we can easily sum the coordinates of two vectors

[5]see Sect. 2.2.3

To determine the translator T_2 we need to know a vector of translation. This vector can be computed as a difference between Euclidean parts of P_{j_2} and $P_{r_2} = R_2 P_{j_2} \tilde{R}_2$, which is a point rotated from P_{j_2} by the rotor R_2. Therefore the vector of translation is

$$t_2 = c^{-1}(P_{r_2}) - c^{-1}(P_{j_2})$$

and the corresponding translator is

$$T_2 = 1 - \frac{1}{2} t_2 e_\infty.$$

A construction of the rotor corresponding to the rotation of φ_3 is very similar to R_2. At first we compute the axis as

$$L_3 = P_{j_2} \wedge P_3 \wedge e_\infty,$$

where $P_3 = c(j_2 + e_2)$. The rotor is then expressed as

$$R_3 = \cos\frac{\varphi_3}{2} + L_3^* \sin\frac{\varphi_3}{2}.$$

The angle φ_4 provides a rotation that affects the endpoint P_{j_4} only in the case of non-zero φ_5. It in fact rotates the link l_{34}. The axis of rotation is a line passing through the points P_{j_3}, P_{j_4}, i.e.

$$L_4^0 = P_{j_3} \wedge P_{j_4} \wedge e_\infty,$$

however for the computation of the rotor, we require the axis to be normalized. In CGA for any geometric object O listed in Table 2.4 we can compute a norm as $|O| = \sqrt{O \cdot \tilde{O}}$. Since the distance between points P_{j_3} and P_{j_4} is not equal to 1, the norm of L_4^0 is not equal to 1 either. Therefore the axis of rotation for the rotor R_4 is

$$L_4 = \frac{L_4^0}{\sqrt{L_4^0 \cdot \tilde{L_4^0}}}$$

and the corresponding rotor is

$$R_4 = \cos\frac{\varphi_4}{2} + L_4^* \sin\frac{\varphi_4}{2}.$$

The axis of the last rotation is expressed as

$$L_5 = P_{j_3} \wedge P_5 \wedge e_\infty,$$

where $P_5 = c(j_3 + e_2)$ and the rotor is computed as

$$R_5 = \cos\frac{\varphi_5}{2} + L_5^* \sin\frac{\varphi_5}{2}.$$

Now the final position of the end point denoted as $\bar{P}_{j_4}$ can be computed as

$$\bar{P}_{j_4} = (R_1 T_2 R_3 R_4 R_5) P_{j_4} (R_1 T_2 R_3 R_4 R_5)\tilde{}.$$

Let us note that in CGA it does not matter if we apply transformations one by one or if we apply one composed transformation. In order to provide a chain for computation of the final position of any joint, let us denote transformations in a different way. Let V_i be the versor that realizes the last transformation in a chain needed for the computation of the final position of P_{j_i}, i.e. this transformation does not have any effect on the points $P_{j_1}, \dots, P_{j_{i-1}}$. Therefore $V_4 = R_4 R_5$, since both rotors only affect a position of the point P_{j_4}. Then we can express the computation of any point as

$$\begin{aligned}
\bar{P}_{j_1} &= V_1 P_{j_1} \tilde{V}_1, \\
\bar{P}_{j_2} &= (V_1 V_2) P_{j_2} (V_1 V_2)^{\sim}, \\
\bar{P}_{j_3} &= (V_1 V_2 V_3) P_{j_2} (V_1 V_2 V_3)^{\sim}, \\
\bar{P}_{j_4} &= (V_1 V_2 V_3 V_4) P_{j_4} (V_1 V_2 V_3 V_4)^{\sim},
\end{aligned}$$

where versors in equations above are defined as

$$\begin{aligned}
V_1 &:= R_1, \\
V_2 &:= \begin{cases} R_2, & \text{for } \bar{P}_{j_2}, \\ T_2, & \text{for } \bar{P}_{j_3}, \bar{P}_{j_4}, \end{cases} \\
V_3 &:= R_3, \\
V_4 &:= R_4 R_5.
\end{aligned}$$

8.3 MATLAB® TOOLBOX IMPLEMENTATION

Here, we implement the algorithm described above using the Clifford Multivector Toolbox for MATLAB. Let us note that `einf`, `eo` in the implementation stand for predefined constants $e_4 + e_5, 0.5(e_5 - e_4)$, respectively, which are not implemented in the toolbox by default. For simplicity we also define a function called `CreatePoint`.

Listing 8.1: CreatePoint function

```
1 function res = CreatePoint(x,y,z)
2 res = x*e1+y*e2+z*e3 + 0.5*(x*e1+y*e2+z*e3)^2*einf + eo;
3 end
```

The algorithm is realized as the MATLAB's function `Forw_Kin` with inputs ang1,..., ang5 corresponding to angles $\varphi_1, \dots \varphi_5$. At the beginning of the code we have to initialize a particular algebra, in our case $\mathbb{G}_{4,1}$ by command `clifford_signature(4,1)`. In the code we can see default functions such as `dual()`, `wedge()`, `reverse()`, `scalar_product()` (for elements of the same grade it has the same functionality as inner product) whose meaning is clear. For the geometric product we use the symbol `*` and the function `part(mult,coeff)` returns coefficients of a particular basis blade. As an output we get the Euclidean x, y, z coordinates of the position of the endpoint.

Listing 8.2: Forw˙Kin function by the Clifford Toolbox

```
1  function [Pj4_f_x, Pj4_f_y, Pj4_f_z] = ...
2  Forw_Kin(ang1,ang2,ang3,ang4,ang5)
3
4  clifford_signature(4,1);
5  d1 = 200; d2 = 680; d3 = 150; d4 = 880; l12 = 890; l34 = 254
6
7  Pj1_x=d1;               Pj1_y=0;          Pj1_z=d2;
8  Pj2_x=d1;               Pj2_y=0;          Pj2_z=d2+l12;
9  Pj3_x=d1+d4;            Pj3_y=0;          Pj3_z=d2+l12+d3;
10 Pj4_x=d1+d4+l34;        Pj4_y=0;          Pj4_z=d2+l12+d3;
11
12 Pj1 = createPoint(Pj1_x, Pj1_y, Pj1_z);
13 Pj2 = createPoint(Pj2_x, Pj2_y, Pj2_z);
14 Pj3 = createPoint(Pj3_x, Pj3_y, Pj3_z);
15 Pj4 = createPoint(Pj4_x, Pj4_y, Pj4_z);
16
17 %Transformation 1
18 P1_help = createPoint(0,0,1);
19 L1      = dual(wedge(eo,P1_help,einf));
20 R1      = cos(ang1/2) + L1*sin(ang1/2); % the rotor of ang1
21
22 %Transformation 2
23 P2_help = createPoint(Pj1_x, Pj1_y+1, Pj1_z);
24 L2      = dual(wedge(Pj1,P2_help,einf));
25 R2      = cos(ang2/2) + L2*sin(ang2/2); % the rotor of ang2
26
27 Pj2_rot = R2 * Pj2 * reverse(R2);
28 t2      = (part(Pj2_rot,2)-Pj2_x)*e1 +...
29           (part(Pj2_rot,3)-Pj2_y)*e2 +...
30           (part(Pj2_rot,4)-Pj2_z)*e3;
31 T2      = 1 - 0.5*t2*einf; % the translator of ang2
32
33 %Transformation 3
34 P3_help = createPoint(Pj2_x, Pj2_y + 1, Pj2_z);
35 L3      = dual(wedge(Pj2,P3_help,einf));
36 R3      = cos(ang3/2) + L3*sin(ang3/2); % the rotor of ang3
37
38 %Transformation 4
39 L4_     = dual(wedge(Pj3,Pj4,einf));
40 L4      = L4_/(sqrt(scalar_product(L4_,reverse(L4_))));
41 R4      = cos(ang4/2) + L4*sin(ang4/2); % the rotor of ang4
42
43 %Transformation 5
44 P5_help = createPoint(Pj3_x, Pj3_y + 1, Pj3_z);
45 L5      = dual(wedge(Pj3,P5_help,einf));
46 R5      = cos(ang5/2) + L5*sin(ang5/2); % the rotor of ang5
47
48 %Outputs
```

```
49 Pj4_f = (R1*T2*R3*R4*R5) * Pj4 * reverse(R1*T2*R3*R4*R5);
50
51 Pj4_f_x = part(Pj4_f ,2);
52 Pj4_f_y = part(Pj4_f ,3);
53 Pj4_f_z = part(Pj4_f ,4);
54 end
```

8.4 THE GAALOP IMPLEMENTATION

The following listing shows the GAALOPScript for the kinematics algorithm of Sect. 8.2:

Listing 8.3: GAALOPScript

```
1  expp = {
2  1 + _P(1) + _P(1)*_P(1)/2 + _P(1)*_P(1)*_P(1)/6 +
3  _P(1)*_P(1)*_P(1)*_P(1)/24
4  }
5
6  d1 = 200; d2 = 680; d3 = 150; d4 = 880; l12 = 890; l34 = 254;
7
8  Pj1_x=d1; Pj1_y=0; Pj1_z=d2;
9  Pj2_x=d1; Pj2_y=0; Pj2_z=d2+l12;
10 Pj3_x=d1+d4; Pj3_y=0; Pj3_z=d2+l12+d3;
11 Pj4_x=d1+d4+l34; Pj4_y=0; Pj4_z=d2+l12+d3;
12
13 Pj1 = createPoint(Pj1_x, Pj1_y, Pj1_z);
14 Pj2 = createPoint(Pj2_x, Pj2_y, Pj2_z);
15 Pj3 = createPoint(Pj3_x, Pj3_y, Pj3_z);
16 Pj4 = createPoint(Pj4_x, Pj4_y, Pj4_z);
17
18 //Transformation 1
19 P1_help = createPoint(0,0,1);
20 ?L1 = *(e0^P1_help^einf);
21 ?R1 = expp(ang1*L1/2);
22
23 // Transformation 2
24 ?P2_help = createPoint(Pj1_x, Pj1_y+1, Pj1_z);
25 ?L2 = *(Pj1^P2_help^einf);
26 ?R2 = expp(ang2*L2/2);
27
28 ?Pj2_rot = R2 * Pj2 * ~R2;
29 ?t2 = (Pj2_rot.e1-Pj2_x)*e1 + (Pj2_rot.e2-Pj2_y)*e2
30 +(Pj2_rot.e3-Pj2_z)*e3;
31 ?T2 = expp(-0.5*t2*einf);
32
33 // Transformation 3
34 P3_help = createPoint(Pj2_x, Pj2_y + 1, Pj2_z);
35 ?L3 = *(Pj2^P3_help^einf);
```

```
36 ?R3 = expp(ang3*L3/2);
37
38 // Transformation 4
39 ?L4_ = *(Pj3^Pj4^einf);
40 ?L4 = L4_/abs(L4_);
41 ?R4 = expp(ang4*L4/2);
42
43 // Transformation 5
44 ?P5_help = createPoint(Pj3_x, Pj3_y + 1, Pj3_z);
45 ?L5 = *(Pj3^P5_help^einf);
46 ?R5 =  expp(ang5*L5/2);
47
48 // Outputs
49 ?Pj4_f = (R1*T2*R3*R4*R5) * Pj4 * ~(R1*T2*R3*R4*R5);
```

When comparing this code with the Clifford Toolbox code of the previous section we can see: there are many similarities in these two descriptions of Geometric Algebra algorithms. They only differ in:

in GAALOPScript the transformations are described by the macro[6] *expp* based on the Taylor series

instead of the compact form with the ∗ and ˜ signs, the Clifford Toolbox uses the functions dual() and reverse()

the part function of the MATLAB Toolbox has to be expressed algebraically in GAALOPScript

the question marks in front of many multivectors

Mainly these are only some simple syntax differences, but what about the question marks? The question marks in front of some multivectors indicate that these multivectors should be computed explicitly by GAALOP (further improvements can be achieved with exclamations marks according to Sect. 5.2). All other multivectors are treated as intermediate auxiliary results. The result is optimized MATLAB code according to the following MATLAB function:

Listing 8.4: Generated optimized MATLAB code

```
1  function [L1, L2, L3, L4, L4_, L5, P2_help, P5_help,
2  Pj2_rot, Pj4_f, R1, R2, R3, R4, R5, T2, t2]
3  = Forw_Kin (ang1, ang2, ang3, ang4, ang5)
4  L1(7) = -1.0; % e1 ^ e2
5  R1(1) = 0.00260416666666666 * power(ang1,4.0)
6  - 0.125 * ang1 * ang1 + 1.0; % 1.0
7  R1(7) = 0.0208333333333333 * ang1 * ang1 * ang1
8  - ang1/2.0; % e1 ^ e2
9  P2_help(2) = 200.0; % e1
10 P2_help(3) = 1.0; % e2
11 P2_help(4) = 680.0; % e3
```

[6]see Sect. 3.2.2

```
12 P2_help(5) = 251200.5; % einf
13 P2_help(6) = 1.0; % e0
14 L2(8) = 1.0; % e1 ^ e3
15 L2(9) = 680.0; % e1 ^ einf
16 L2(14) = -200.0; % e3 ^ einf
17 R2(1) = 0.00260416666666666 * power(ang2,4.0)
18 - 0.125 * ang2 * ang2 + 1.0; % 1.0
19 R2(8) = ang2 / 2.0 - 0.0208333333333333
20 * ang2 * ang2 * ang2; % e1 ^ e3
21 R2(9) = 340.0 * ang2 - 14.1666666666666
22 * ang2 * ang2 * ang2; % e1 ^ einf
23 R2(14) = 4.16666666666666 * ang2 * ang2 * ang2
24 - 100.0 * ang2; % e3 ^ einf
25 Pj2_rot(2) = -2.0 * R2(1) * R2(9) - 200.0 * R2(8) * R2(8)
26 + (3140.0 * R2(1) - 2.0 * R2(14)) * R2(8)
27 + 200.0 * R2(1) * R2(1); % e1
28
29 ...
30
31 end
```

The parameters of this function are the angles ang1,...,ang5. All the multivectors to be explicitly computed are described by the computations of all its non-zero coefficients. The multivector L1, for instance, is described as an array with one non-zero entry. As indicated by the comment, this entry (with index 7) is related to the basis blade $e_1 \wedge e_2$.

8.5 GAALOPWEB FOR MATLAB®

In order to introduce GAALOP to the MATLAB community, our requirement was to extend GAALOP in a way to make it easy to handle for MATLAB users. We developed GAALOPWeb also for MATLAB. The big advantage of the web-based solution is that it can be used from everywhere without any need of installing a software.

There are two ways for MATLAB users to deal with GAALOPWeb. One is simply using the web interface of Sect. 4.1.

Figure 8.2 shows a screenshot of *GAALOPWeb for MATLAB*[7] with the GAALOPScript of Listing 8.3[8]. By pushing the Run button, optimized MATLAB code is generated according to Listing 8.4. You are able to use the optimized code in your MATLAB environment based on copy and paste.

For users who would like to have a solution contained within the MATLAB environment there is a second way

- download the *MATLAB Connector For GAALOPWeb* from the web page (see Fig. 8.2),

[7]www.gaalop.de/gaalopweb

[8]in this example visualization is not used

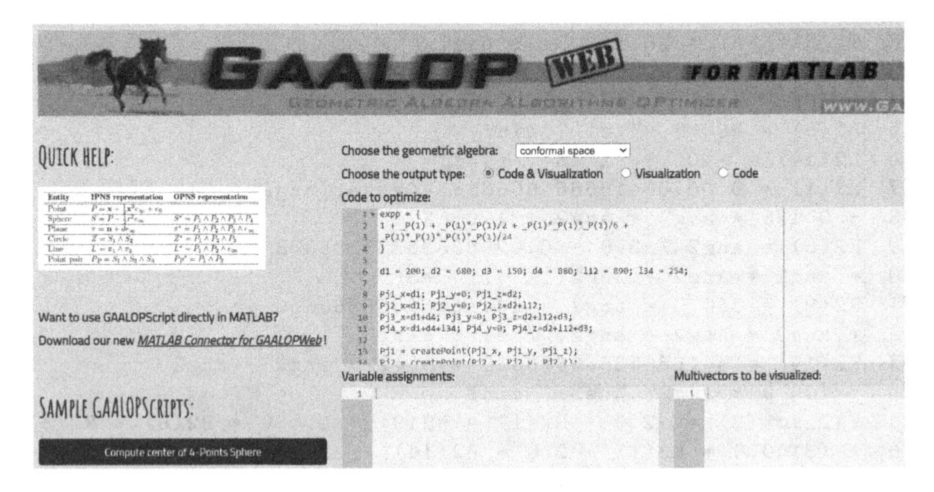

Figure 8.2 The screen of GAALOPWeb for MATLAB.

edit a GAALOPScript as a text file in MATLAB[9] and use the MATLAB function *gaalopscriptFile2Matlab* **or**

use the MATLAB function *gaalopscriptCellArr2Matlab* with this GAALOPScript as parameter in order to automatically generate an optimized MATLAB function.

the optimized MATLAB function according to Listing 8.4 will be automatically copied into your MATLAB environment.

8.6 COMPARISON OF RUN-TIME PERFORMANCE

Let us demonstrate the functionality and improvement of the kinematics algorithm by optimizing the code with using GAALOP/GAALOPWeb. The Fig. 8.3 displays graphs of the same motion of the robot, one obtained from values computed in MATLAB via our optimized code and the other graph from ABB Robot Studio by built-in computational software. By motion we understand a sequence of multiple inputs between two configurations of the robot. Let us denote that in MATLAB we chose the same values of time as ABB Robot Studio generated. These values are influenced by speed of the motion which is set in ABB Robot Studio software and which is not dependent on the computation time. By the optimized code we obtained the same output values as by ABB Robot Studio, so the graphs are identical.

To measure a computation time of the original and optimized codes we use MATLAB's built-in function `timeit()` which runs the code 100 times and then it returns the expected value. The results of our tests (we measured each code 3 times) are displayed in Table 8.1. The improvement of the runtime is significant, the time

[9]in this case according to Listing 8.3.

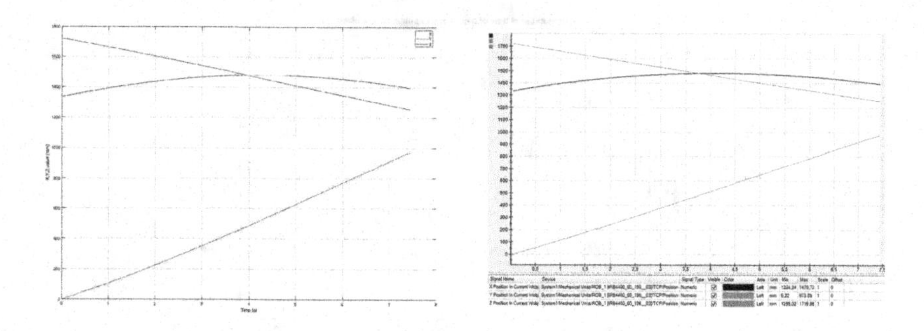

Figure 8.3 Motion of the robot: left – MATLAB-optimized code, right – ABB Robot Studio. This shows that the functionality of both algorithms is identical.

to compute the output with the original Multivector Toolbox code is around $0.10s$ and the time after optimization is around $0.0001s$. Therefore the optimized code is approximately 1000 times faster.

TABLE 8.1 Computation Time of Algorithms in (s)

MATLAB toolbox results	Optimized code
0.1258	0.00008
0.1026	0.00009
0.0989	0.00009

In Fig. 8.4 we display a graph of computation time against the number of steps in a motion of the robot. With the optimization we are able to compute very smooth motions (a sequence of many steps between two configurations) in very short time which would not be possible with the original code. For all tests and computations we used a computer with parameters described in Table 8.2.

TABLE 8.2 Computational Computer Specification

Processor	Intel(R) Core(TM) i7-2600 CPU @ 3.40 GHz 3.70 GHz
Ram	8.00 GB
Windows	64bit system
MATLAB version	9.5.0.1033004 (R2018b) Update 2

As a conclusion, we could demonstrate the power of the Geometric Algebra Computing technology in terms of the expressiveness of the mathematical language and the high runtime performance. The optimized MATLAB code is approximately 1000 times faster than the library-based solution. Based on this code we are able to control the movement of the robot very smoothly. We used MATLAB because of its popularity in the engineering and research community. But, with GAALOP we are not restricted to this programming language: with the conventional GAALOP

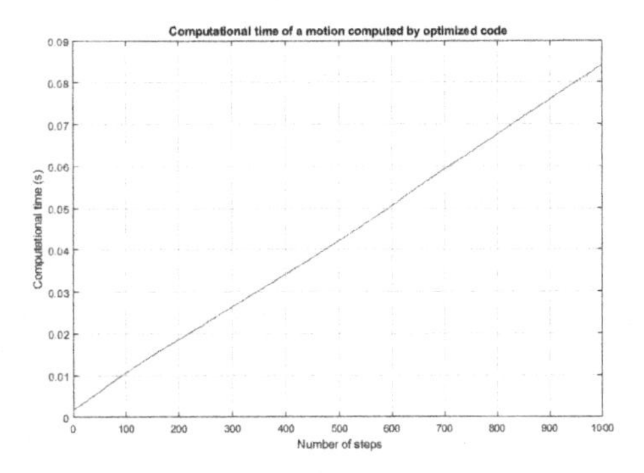

Figure 8.4 Measuring of the runtime through the whole robot's motion.

software we are able to directly transform a MATLAB solution to another programming language. So it is very efficient to develop and test an algorithm in MATLAB and if the results are satisfying, then it is easy, for instance, to switch to C/C++ for an implementation into the robot.

Sections 8.3 and 8.4 showed, that GAALOPScript and the syntax used in the Clifford Multivector Toolbox are already very similar. We intend to extend GAALOPScript in order to make it even easier to switch between these two tools. One approach may be to create macros with a name, that is equal to the function names of the Clifford Multivector Toolbox. This allows a direct usage of the whole code. Nevertheless question marks have to be inserted to indicate the multivectors to be computed explicitly. Since the actual distribution of question marks is very important for the runtime performance of the generated code, research is needed in order to find the best choice.

CHAPTER 9

The Power of High-Dimensional Geometric Algebras

CONTENTS

So far, we used low-dimensional Geometric Algebras up to a dimension of 5. With Conformal Geometric Algebra we are able to handle geometric objects such as spheres and planes. With higher-dimensional Geometric Algebras we are able to handle more complex geometric objects such as conics or quadrics.

The GAC G(5,3) as presented in Chapt. 10, for instance, is an algebra of conics. It is an extension of 2D Compass-Ruler Algebra CRA[1] for geometric objects such as ellipses, parabolas and hyperbolas. Double Conformal Geometric Algebra as an algebra for objects such as quadrics, tori and cyclides is treated in Chapt. 11.

Cubic Conformal Geometric Algebra [38] is presented in Chapt. 12.

9.1 GAALOP DEFINITION

GAALOP/GAALOPWeb already include many predefined Geometric Algebras. The stand-alone GAALOP can even be extended by user-defined ones. The definition of new algebras needs two files: the definition.csv and macros.clu. Here, we describe these files based on the Conformal Geometric Algebra of Sect. 2.2. In order to compute blade products at least two informations are needed: the basis of the Geometric Algebra and the signature of the Geometric Algebra. When using basis vectors with geometric meaning, the usage of another basis is more convenient. In Conformal Geometric Algebra, the standard basis is e_1, e_2, e_3, e_+, e_- with the signatures $e_1^2 = 1, e_2^2 = 1, e_3^2 = 1, e_+^2 = 1, e_-^2 = -1$. But, using the basis $e_1, e_2, e_3, e_\infty, e_0$ is

[1] see Sect. 2.3

DOI: 10.1201/9781003139003-9

geometrically more intuitive, because, as described in Sect. 2.2, e_∞ can be interpreted as infinity and e_0 can be interpreted as the origin.

Making it as easy as possible for the user to integrate his own Geometric Algebra in GAALOP, not only the original basis and the signature of the Geometric Algebra must be given as Geometric Algebra definition, but also another basis with the transformation equations between the first basis and the second basis can be given. In the example of the Conformal Geometric Algebra, this leads to the following Geometric Algebra definition[2]:

- standard basis:e_1, e_2, e_3, e_+, e_-

- used basis: $e_1, e_2, e_3, e_\infty, e_0$

- signature: $e_1^2 = 1, e_2^2 = 1, e_3^2 = 1, e_+^2 = 1, e_-^2 = -1$

- transformations from standard basis to the used basis:

$e_+ = 0.5 * e_\infty - e_0$ and $e_- = 0.5 * e_\infty + e_0$

- transformations from used basis to standard basis:

$e_0 = 0.5 * e_- - 0.5 * e_+$ and $e_\infty = e_- + e_+$

This is defined in the Geometric Algebra definition file definition.csv according to Listing 9.1.

Listing 9.1: Definition.csv for Conformal Geometric Algebra.

```
1  1,e1,e2,e3,einf,e0
2  ep=0.5*einf-e0,em=0.5*einf+e0
3  1,e1,e2,e3,ep,em
4  e1=1,e2=1,e3=1,ep=1,em=-1
5  e0=0.5*em-0.5*ep,einf=em+ep
```

Definition.csv is a human-readable comma-separated-value file, that contains five lines:

1. comma-separated list of the used basis vectors, starting with the scalar.
2. comma-separated list of the mappings from the standard basis vectors to the used basis vectors
3. comma-separated list of the standard basis vectors, starting with the scalar.
4. comma-separated list of the signature of the standard basis vectors
5. comma-separated list of the mappings from the used basis vectors to the standard basis vectors

[2]see [62, 63]

The CSV-format is chosen because it is very compact. The file macros.clu according to Listing 9.2 defines all the predefined functions of a specific algebra and needs at least the createPoint() and the Dual() functionaliy.

Listing 9.2: Basic macros of Conformal Geometric Algebra.

```
1  createPoint = {
2     _P(1)*e1+_P(2)*e2+_P(3)*e3
3      +0.5*(_P(1)*_P(1)+_P(2)*_P(2)+_P(3)*_P(3) )*einf+e0
4  }
5
6  Dual = {
7     _P(1)*(-e1^e2^e3^einf^e0)
8  }
```

createPoint(x,y,z) computes a conformal point based on the 3D coordinates x, y and z according to Table 2.4. Please notice that the parameters of macros are denoted by `_P(1), _P(2)` ...

In principle, the dual of a multivector is computed using the division with the pseudoscalar[3]. Since, in this case, the norm of the pseudoscalar is 1, the multiplication with the pseudoscalar can be used.

9.2 VISUALIZATION

GAALOPWeb uses the Ganja visualization as described in Sect. 4.3 and used in the previous sections. The stand-alone GAALOP offers two visualization possibilities according to the master thesis [63]:

1. Vis2d for the 2D Compass Ruler Algebra (see Sect. 2.3). Vis2d interprets the particular CRA blades and renders them,
2. Vis3d for higher-dimensional Geometric Algebras (such as the ones of the following three chapters).

After the definition of a new algebra according to the previous section, the Vis3d visualization can be used immediately for its examination. Since Vis3d always renders three-dimensional, Geometric Algebras in the plane – such as the Geometric Algebra for Conics according to Chapt. 10 – curves are rendered as cylinders into the z-axis. Please refer to [63] for details about the GAALOP visualizations.

In the following, the Geometric Algebra for Conics according to Chapt. 10 is presented based on the GAALOPWeb visualization and the Chapt. 11 shows the Double Conformal Geometric Algebra based on the Vis3d visualizer of GAALOP.

[3] see Sect. 2.1

CHAPTER 10

GAALOPWeb for Conics

CONTENTS

The Geometric Algebra for Conics (GAC) as a G(5,3) is originally based on the work of Christian Perwass in his book [57]. In the meantime, this work is continued in the paper: Hrdina, Navrat, Vasik – "Geometric Algebra for Conics" [43]. This paper explains also how subalgebras such as the Zamora 2D QGA G(4,2) [68] can be used within GAC as a subalgebra.

10.1 GAALOP DEFINITION

As described in Chapt. 9, a definition of a Geometric Algebra within GAALOP (or GAALOPWeb) needs the two specific files definition.csv and macros.clu.

10.1.1 definition.csv

Listing 10.1 shows the file definition.csv.

Listing 10.1: Definition.csv for the GAC.

```
1  1,e1,e2,e0p,e0m,e0c,einfp,einfm,einfc
2  ep1=0.5*einfp-e0p,em1=0.5*einfp+e0p,ep2=0.5*einfm-e0m,
     em2=0.5*einfm+e0m,ep3=0.5*einfc-e0c,em3=0.5*einfc+e0c
3  1,e1,e2,ep1,ep2,ep3,em1,em2,em3
4  e1=1,e2=1,ep1=1,ep2=1,ep3=1,em1=-1,em2=-1,em3=-1
5  e0p=0.5*em1-0.5*ep1,einfp=em1+ep1,e0m=0.5*em2-0.5*ep2,
     einfm=em2+ep2,e0c=0.5*em3-0.5*ep3,einfc=em3+ep3
```

The first line shows the scalar and the 8 basis elements of GAC according to Chapt. 6 of [43] with a Maple implementation of GAC. $e1, e2$ are the usual basis

DOI: 10.1201/9781003139003-10

vectors of the plane while $e0p, e0m, e0$ describe the origin and einfp, einfm, einfc describe infinitiy. The underlying standard basis vectors together with their signatures are described in the lines 3 and 4. Line 2 as well as line 5 show the transformations between these basis vectors. Please notice that the indented lines belong to the preceding lines.

10.1.2 macros.clu

Listing 10.2 shows the file macros.clu for GAC starting with the two macros needed for every Geometric Algebra: createPoint() and Dual().

Listing 10.2: Macros.clu for the GAC: basic macros.

```
1  // GAC macros.clu
2
3  createPoint = {
4      e0p + _P(1)*e1 + _P(2)*e2
5    + 0.5*(_P(1)*_P(1)+_P(2)*_P(2))*einf
6    + 0.5*(_P(1)*_P(1)-_P(2)*_P(2))*einfm
7    + _P(1)*_P(2)*einfc+_P(3)*0
8  }
9
10 // Define the dual operator * that dualizes A as (*A).
11 Dual = {
12   _P(1)/(e0p^e0m^e0c^e1^e2^einfp^einfm^einfc)
13 }
14
15 DualIO = { _P(1).(e0p^e1^e2^einfp^einfm^einfc) }
16 DualOI = { _P(1).(e0p^e0m^e0c^e1^e2^einfp) }
```

The macro createPoint() is defined according to Eq. (2) of [43][1]. The macro dual() is defined by the division by the pseudoscalar while also the macros DualIO() and DualOI() can be used.

DualIO is the duality operator from GAC IPNS 1-blade entities to GAC OPNS 5-blade entities (the wedge of 5 points). Only 5 points are required to define any quadratic plane curve. The full GAC OPNS entities are 7-blades by the Dual *, but they all have the factor `e0m^e0c`,which can be abridged (removed). DualIO performs this dualization to the abridged 5-blade form, that also serve as primary OPNS entities. The DualIO pseudoscalar `Iio=(e0p^e1^e2^einfp^einfm^einfc)` removes `e0m^e0c`, and produces an OPNS 5-blade entity without these factors.

DualOI is the pseudoinverse operation to the DualIO operation, using the reciprocal 5-blade pseudoscalar `Ioi = (e0p^e0m^e0c^e1^e2^einfp)` and `Iio.Ioi = 1`. So, DualOI takes a primary GAC OPNS entity, up to a 5-blade, and produces the corresponding GAC IPNS entity.

[1]In GAALOP, we have to define createPoint with 3 parameters (x,y,z), but only really use (x,y)

In Proposition 4.2 (page 8 of [43]), the GAC OPNS 5-blade entities are:

- Wedge of 5 GAC points = GAC OPNS entity for general conic plane curve.

- Wedge of 4 GAC points and einfc = GAC OPNS entity for axes-aligned conic.

- Wedge of 3 GAC points and `einfm^einfc` = GAC OPNS circle of GAC points. Note: wedge of 3 CGA points (in CGA subalgebra) is circle of CGA points.

- Wedge of 2 GAC points and `einfp^einfm^einfc` = GAC OPNS line.

All of these 5-blades dualize via DualOI to GAC IPNS 1-blade entities.

Besides these basic definitions, macros.clu also provides functionality for the predefinition of geometric objects and transformations according to the following sections.

10.2 GAC OBJECTS

Listing 10.3 shows the extension of macros.clu for the predefinition of GAC objects.

Listing 10.3: Macros.clu for the GAC: geometric objects.

```
1  // GAC macros.clu
2
3  // (1 semi-diameter a, 2: semi-diameter b)
4  Ellipse = {
5   (_P(1)*_P(1) + _P(2)*_P(2))*e0p -
6   (_P(1)*_P(1) - _P(2)*_P(2))*e0m -
7   (_P(1)*_P(1)*_P(2)*_P(2))*einfp
8  }
9
10 // x, y, r
11 Circle = {
12  PNT = _P(1)*e1 + _P(2)*e2
13        + (1/2)*(_P(1)*_P(1) + _P(2)*_P(2))*einfp + e0p;
14  PNT - (1/2)*_P(3)*_P(3)*einfp
15 }
16
17 // semi-diameters a, b
18 Hyperbola = {
19  (_P(1)*_P(1) - _P(2)*_P(2))*e0p -
20  (_P(1)*_P(1) + _P(2)*_P(2))*e0m +
21  (_P(1)*_P(1)*_P(2)*_P(2))*einfp
22 }
23
24 // (p semi-latus rectum), half the diameter
25 // through focus parallel to directrix
26 Parabola = {
```

```
27 |    e0p + e0m + _P(1)*e2
28 |}
```

An ellipse centred at the origin is defined as Ellipse(a,b) with the semi-diameters a and b according to Eq. (6) of [43][2].

A circle as a specific ellipse is defind as Circle(x,y,r) with center point (x,y) and radius r according to proposition 3.4 of [43][3].

A hyperbola is defined as Hyperbola(a,b) with the semi-diameters a and b according to Eq. (7) of [43]. A parabola is defined by Parabola(p) with semi-latis rectum p according to Eq. (8) of [43].

Please notice that the propositions 3.2–3.5 of [43] give IPNS entities in general positions.

Figure 10.1 shows the definition of an ellipse E with the semi-diameters a and b to be visualized using GAALOPWeb. Since the parameters a and b are not defined,

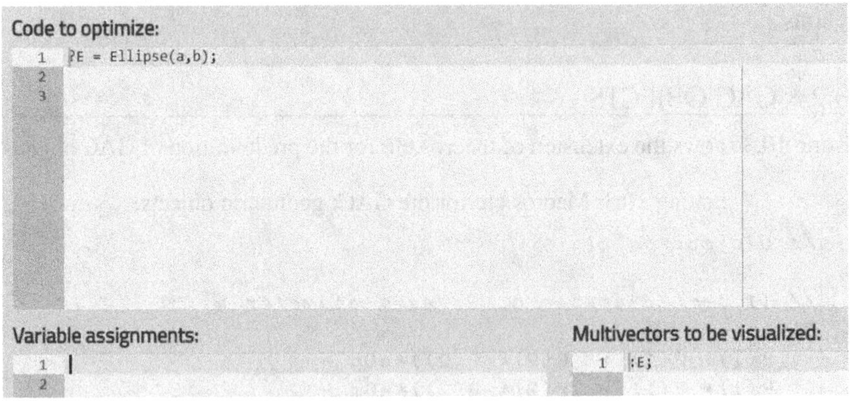

Figure 10.1 An ellipse to be visualized.

they can be adjusted using sliders as shown in Fig. 10.2.

Listing 10.4 shows the resulting Python code.

Listing 10.4: resulting Python code for an ellipse.

```
1  import numpy as np
2
3  def ellipse(a, b):
4      E = np.zeros(256)
5      E[3] = ((-(a * a)) - 0.5) * b * b - (a * a) / 2.0; # ep1
6      E[4] = (a * a) / 2.0 - (b * b) / 2.0; # ep2
7      E[6] = (0.5 - a * a) * b * b + (a * a) / 2.0; # em1
8      E[7] = (b * b) / 2.0 - (a * a) / 2.0; # em2
9      return E
```

[2] one change from the first + to a − has to be made to get the correct ellipse shape.

[3] The GAC circle has the same form as the CGA circle. An Ellipse with a=b=r and a Circle(0,0,r) are equal up to scale

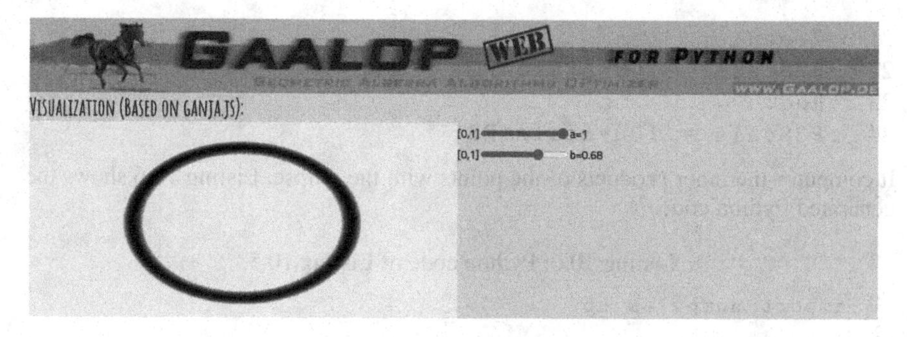

Figure 10.2 The visualization of an ellipse with GAALOPWeb.

The visualization is done by the Ganja visualizer for high-dimensional algebras as an implicit root finder. That is, it visualizes a boundary around the actual null space to find this boundary. It considers the outer products Euclidean norm, and renders where this norm becomes small enough. A different line width can be reached by simply scaling the multivector to be visualized.

For that we extend the definition of the multivector to be visualized in the following GAALOPScript to a slider for scaling values for instance for [0..5] by `:EScale = 5*scale*E;`. It computes the ellipse with $a = 1.5$ and $b = 0.8$ as well as the 3 points $P1, P2$ and $P3$ and the general point GP (with the general coordinates x and y).

Listing 10.5: Points on, inside or outside of an ellipse.

```
1  // Code to optimize:
2  E = Ellipse(1.5,0.8);
3
4  x1=1.5; y1=0;
5  x2=1.7; y2=0.3;
6  x3=0; y3=0;
7  P1 = createPoint(x1,y1,0);
8  P2 = createPoint(x2,y2,0);
9  P3 = createPoint(x3,y3,0);
10 GP = createPoint(x,y,0);
11
12 ?IPN=GP.E;
13 ?IP1=P1.E;
14 ?IP2=P2.E;
15 ?IP3=P3.E;
16
17 // Multivectors to be visualized:
18 :EScale = 5*scale*E;
19 :Blue;
20 :P1Scale = 100*scale*P1;
```

```
21 :Green;
22 :P2Scale = 100*scale*P2;
23 :Red;
24 :P3Scale = 100*scale*P3;
```

It computes the inner products of the points with the ellipse. Listing 10.6 shows the generated Python code.

Listing 10.6: Python code of Listing 10.5.

```
1  import numpy as np
2
3  def script(x, y):
4          IPN = np.zeros(256)
5          IPN[0] = (-(4.0 * y * y)) - x * x + 4.0; # 1.0
6          IP2 = np.zeros(256)
7          IP2[0] = -5.0; # 1.0
8          IP3 = np.zeros(256)
9          IP3[0] = 4.0; # 1.0
10         return IP1, IP2, IP3, IPN
```

The inner product IPN of the ellipse and the general point shows the implicit form of the ellipse depending on the parameters x and y. The inner product IP1 of the ellipse and the point P1 is not computed. This means that it is zero indicating that this point P1 is on the ellipse. The inner products with the points P2 and P3 show different signs indicating that they are on different sides of the ellipse.

The result of Listing 10.5 is shown in Fig. 10.3.

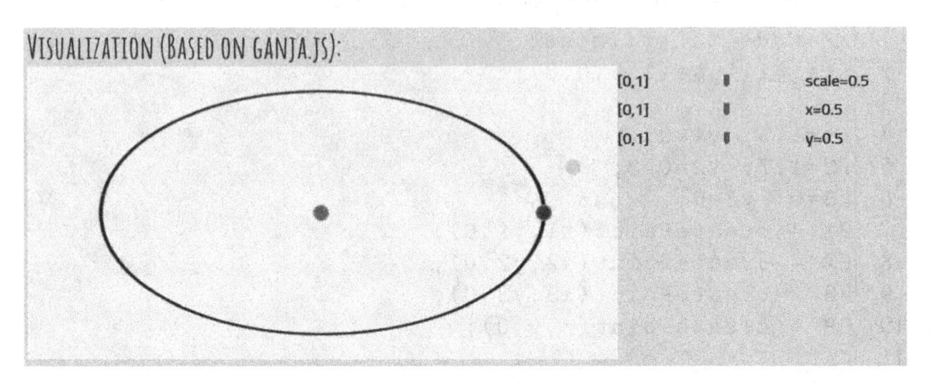

Figure 10.3 The result of Listing 10.5.

Conics can also be defined based on 5 points. Listing 10.7 computes them based on the outer product of four fixed points and one point with variable y-coordinate p2y[4].

[4]the variable scale is used for controling the shape of the geometric objects

Listing 10.7: GAALOPScript for some transformations of an ellipse.

```
1  // Code to optimize:
2  // conics from points
3
4  P1 = createPoint(-0.3, 0,0);
5  P2 = createPoint(-0.2, p2y,0);
6  P3 = createPoint(-0.1, 0.3,0);
7  P4 = createPoint( 0.3,-0.1,0);
8  P5 = createPoint( 0.2,-0.3, 0);
9  ?Conic = DualOI( P1^P2^P3^P4^P5 );
10
11 // Multivectors to be visualized:
12 factor = 500;
13 factor2 = 500000;
14 :Red;
15 :P1V = P1*scale*factor;
16 :Green;
17 :P2V = P2*scale*factor;
18 :Red;
19 :P3V = P3*scale*factor;
20 :P4V = P4*scale*factor;
21 :P5V = P5*scale*factor;
22 :Blue;
23 :ConicV = Conic*scale*factor2;
```

With the standard value for p2y=0.5 the conic spanned by these 5 points are 2 intersecting lines according to the GAALOPWeb visualization of Fig. 10.4. If we change

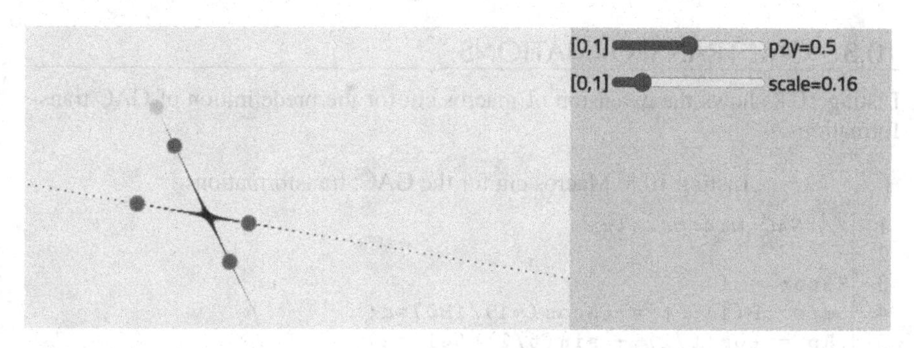

Figure 10.4 GAALOPWeb visualization of two intersecting lines based on 5 points.

the value to p2y=0.35 , for instance, the conic spanned by these 5 points is an ellipse according to the GAALOPWeb visualization of Fig. 10.5.

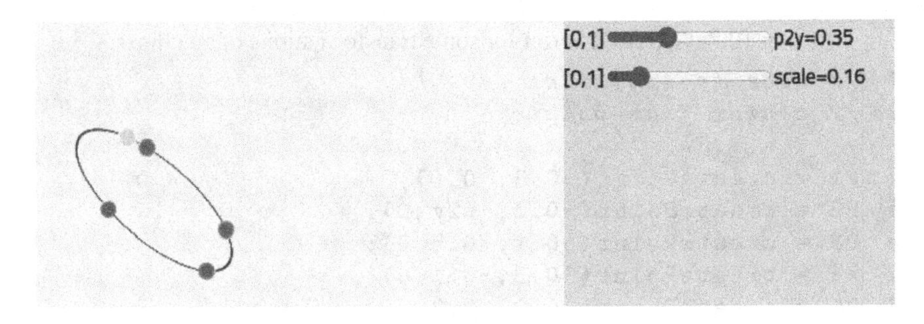

Figure 10.5 GAALOPWeb visualization of an ellipse based on 5 points.

Changing the value to p2y=0.06 results in a hyperbola according to the GAALOPWeb visualization of Fig. 10.6.

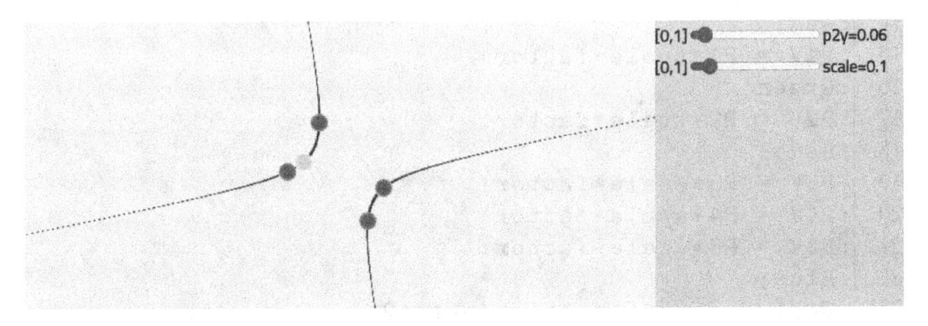

Figure 10.6 GAALOPWeb visualization of a hyperbola based on 5 points.

10.3 GAC TRANSFORMATIONS

Listing 10.8 shows the extension of macros.clu for the predefinition of GAC transformations.

Listing 10.8: Macros.clu for the GAC: transformations.

```
1  // GAC macros.clu
2
3  Rotor = {
4   d = _P(1); t = (acos(-1)/180)*d;
5   Rp = cos(t/2) - sin(t/2)*(e1^e2);
6   R1 = cos(t) - sin(t)*(e0c^einfm);
7   R2 = cos(t) + sin(t)*(e0m^einfc);
8   (R2^R1)*Rp
9  }
10
11 TranslatorX = {
12  x = _P(1);
13  Tp = 1 - (1/2)*x*(e1^einfp);
```

```
14   Tm = 1 - (1/2)*x*(e1^einfm) + (1/4)*x*x*(einfp^einfm);
15   Tc = 1 - (1/2)*x*(e2^einfc);
16   Tp*Tm*Tc
17 }
18
19 TranslatorY = {
20   y = _P(1);
21   Tp = 1 - (1/2)*y*(e2^einfp);
22   Tm = 1 + (1/2)*y*(e2^einfm) - (1/4)*y*y*(einfp^einfm);
23   Tc = 1 - (1/2)*y*(e1^einfc);
24   Tp*Tm*Tc
25 }
26
27 Dilator = {
28   a = _P(1);
29   Sp = (a+1)/(2*sqrt(a)) + (a-1)/(2*sqrt(a))*(e0p^einfp);
30   Sm = (a+1)/(2*sqrt(a)) + (a-1)/(2*sqrt(a))*(e0m^einfm);
31   Sc = (a+1)/(2*sqrt(a)) + (a-1)/(2*sqrt(a))*(e0c^einfc);
32   Sp*Sm*Sc
33 }
```

Rotor(d) describes a rotation around the origin by an angle d in degrees according to proposition 5.1 of [43]. An entity A is transformed as `Rotor(d)*A*~Rotor(d)`. Note, the paper [43] gives a rotor for clockwise rotation, but here we reverse the rotor, as given in the paper [43], into a counter-clockwise rotor.

Translators translate an amount of x in e1-direction or an amount of y in e2-direction according to proposition 5.2 of [43]. The translation of A in e1-direction is expressed by the sandwich product `TranslatorX(x)*A*~TranslatorX(x)` and in e2-direction by
`TranslatorX(y)*A*~TranslatorX(y)`.

Listing 10.9 computes some transformations of an ellipse and visualizes them.

Listing 10.9: GAALOPScript for some transformations of an ellipse.

```
1  // Code to optimize:
2  ?E = Ellipse(0.2,0.5);
3
4  TXE = TranslatorX(0.5)*E*~TranslatorX(0.5);
5  TXYE = TranslatorY(0.1)*TXE*~TranslatorY(0.1);
6  TXYRE = Rotor(45)*TXYE*~Rotor(45);
7
8  // Multivectors to be visualized:
9  :Blue;
10 :EV = E*100;
11 :Red;
12 :TXEV = TXE*100;
13 :Green;
14 :TXYEV = TXYE*100;
```

```
15 :Magenta;
16 :TXYREV = TXYRE*100;
```

Figure 10.7 shows the corresponding GAALOPWeb visualization.

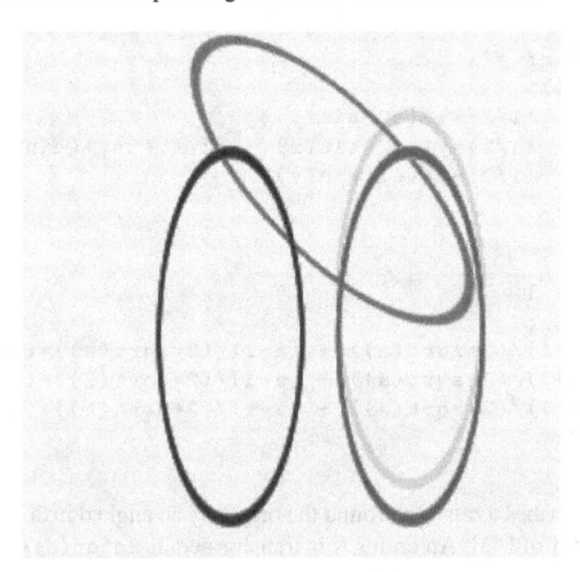

Figure 10.7 GAALOPWeb visualization of some transformations of an ellipse.

Dilator(a) dilates[5] GAC objects by a factor of a>0 according to proposition 5.4 of [43]. For an object A this is done by the sandwich product `Dilator(a)*A*~Dilator(a)`.

Listing 10.10 computes a dilation by 2 of a parabola and an ellipse and visualizes them.

Listing 10.10: GAALOPScript for dilations of conics.

```
1  // Code to optimize:
2  ?C = Circle(0,0,0.5);
3  P = Parabola(1);
4  CD = Dilator(2)*C*~Dilator(2);
5  PD = Dilator(2)*P*~Dilator(2);
6
7  // Multivectors to be visualized:
8  :Green;
9  factor = 10;
10 :CV = factor*C;
11 :Blue;
12 :PV = factor*P;
13 :Red;
```

[5] scales uniformly

```
14 :CDV = factor*CD;
15 :Magenta;
16 :PDV = factor*PD;
```

Figure 10.8 shows the corresponding GAALOPWeb visualization.

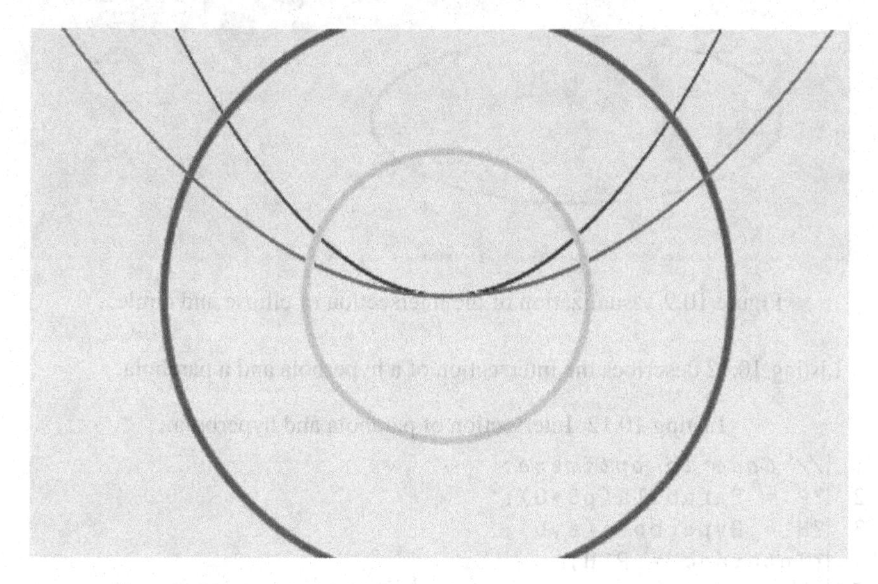

Figure 10.8 GAALOPWeb visualization for dilations of conics.

10.4 INTERSECTIONS

Listing 10.11 describes the intersection of an ellipse with semi-diameter 1.5 and 0.6 and a circle with center at the origin.

Listing 10.11: Intersection of ellipse and circle.

```
1  // Code to optimize:
2  E = Ellipse(1.5,0.6);
3  C = Circle(0,0,radius);
4  ?Intersec = E ^ C;
5
6  // Multivectors to be visualized:
7  :Blue;
8  :EScale = 5*scale*E;
9  :Green;
10 :CScale = 30*scale*C;
11 :Red;
12 :IntersecScale = 3*scale*Intersec;
```

Since its radius is not defined in the GAALOPScript, it can be adjusted by a slider resulting in correspondingly changing point quadruplets as intersection which can be seen in Fig.10.9.

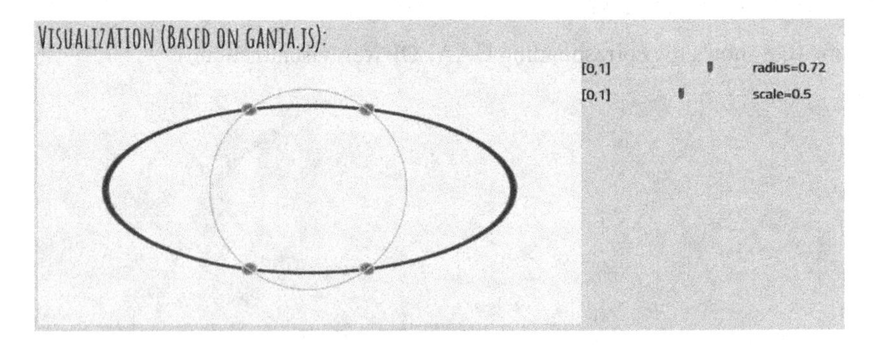

Figure 10.9 Visualization of the intersection of ellipse and circle.

Listing 10.12 describes the intersection of a hyperbola and a parabola.

Listing 10.12: Intersection of parabola and hyperbola.

```
1  // Code to optimize:
2  ?P = Parabola(p8*8);
3  ?H = Hyperbola(a,b);
4  ?Intersec = P^H;
5
6  // Multivectors to be visualized:
7  :Blue;
8  :PScale = 5*scale*P;
9  :Green;
10 :HScale = 40*scale*H;
11 :Red;
12 :IntersecScale = Interscale*10*Intersec;
```

Listing 10.13: generated C code for the intersection of parabola and hyperbola.

```
1  void script(float a, float b, float p8, float Intersec[256]) {
2
3  Intersec[16] = (8.0 * a * a - 8.0 * b * b) * p8; // e2 ^ e0p
4  Intersec[17] = ((-(8.0 * b * b)) - 8.0 * a * a) * p8; // e2 ^ e0m
5  Intersec[19] = 8.0 * a * a * b * b * p8; // e2 ^ einfp
6  Intersec[22] = (-(2.0 * a * a)); // e0p ^ e0m
7  Intersec[24] = a * a * b * b; // e0p ^ einfp
8  Intersec[28] = a * a * b * b; // e0m ^ einfp
9  }
```

The intersecting point pair is visualized in Fig.10.10.

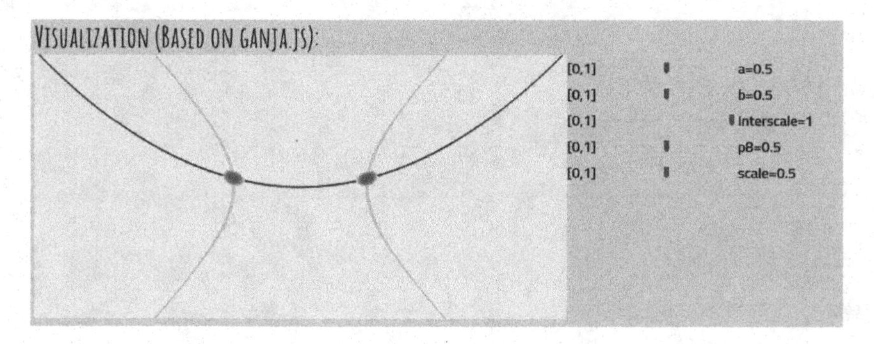

Figure 10.10 Visualization result of Listing 10.12.

CHAPTER 11

Double Conformal Geometric Algebra

CONTENTS

This chapter describes Double Conformal Geometric Algebra (DCGA) [18] based on its integration into GAALOP.

11.1 GAALOP DEFINITION OF DCGA

The integration of DCGA into GAALOP, in principle, is described in [19]. As already mentioned in Chapt. 9, we have to define the files definition.csv and macros.clu. Listing 11.1 shows this definition based on two CGA[1] copies. Since CGA needs 5 basis vectors, DCGA is based on 10 basis vectors.

[1]see Chapt. 2.2

DOI: 10.1201/9781003139003-11

Listing 11.1: Definition.csv for DCGA.

```
1,e1,e2,e3,eo1,ei1,e6,e7,e8,eo2,ei2
e5=eo1+0.5*ei1,e4=0.5*ei1-eo1,e10=eo2+0.5*ei2,e9=0.5*ei2-eo2
1,e1,e2,e3,e4,e5,e6,e7,e8,e9,e10
e1=1,e2=1,e3=1,e4=1,e5=-1,e6=1,e7=1,e8=1,e9=1,e10=-1
eo1=0.5*e5-0.5*e4,eo2=0.5*e10-0.5*e9,ei1=e4+e5,ei2=e9+e10
```

These 10 basis vectors are defined in line 3 together with the scalar 1. Each CGA copy consists of 4 basis vectors squaring to 1 and one basis vector squaring to -1 as defined in line 4. As in CGA, we make a basis transformation in order to be able to use basis vectors with a geometric meaning. The transformations according to the lines 2 and 5 let us use the transformed basis vectors according to line 1: e1, e2, e3 are the basis vectors of the first and e6, e7, e8 of the second CGA copy, eo1 and ei1 represent the origin and infinity of the first and eo2 and ei2 of the second CGA copy. Listing 11.2 shows the basic part of macros.clu.

Listing 11.2: Macros.clu for DCGA: basic macros.

```
{// DCGA} macros.clu
eo = { eo1^eo2 }
ei = { ei1^ei2 }

// Pseudoscalars
IE1 = { e1^e2^e3 }
IE2 = { e6^e7^e8 }
IC1 = { e1^e2^e3^e4^e5 }
IC2 = { e6^e7^e8^e9^e10 }
ID =  { e1^e2^e3^e4^e5^e6^e7^e8^e9^e10 }

// DCGA dualization macro
DD =  { -_P(1).ID() }
// CGA1 and CGA2 dualizations
C1D = { -_P(1).IC1() }
C2D = { -_P(1).IC2() }
// Euclidean 1 and Euclidean 2 dualizations
E1D = { -_P(1).IE1() }
E2D = { -_P(1).IE2() }
// special Dual macro for DCGA dualization operator *
Dual = { DD(_P(1)) }

// Normalize a non-null vector
Normalize = { _P(1)/sqrt(_P(1)._P(1)) }

// CGA1_Point(x,y,z) representing point at (x,y,z)
// in Euclidean 3D space
CGA1_Point = {
 _P(1)*e1 + _P(2)*e2 + _P(3)*e3 +
 (1/2)*(_P(1)*_P(1) + _P(2)*_P(2) + _P(3)*_P(3))*ei1 + eo1
}
// CGA2_Point(x,y,z) representing point at (x,y,z)
// in Euclidean 3D space
CGA2_Point = {
 _P(1)*e6 + _P(2)*e7 + _P(3)*e8 +
 (1/2)*(_P(1)*_P(1) + _P(2)*_P(2) + _P(3)*_P(3))*ei2 + eo2
}
```

```
38  // special DCGA createPoint macro
39  createPoint = {
40    CGA1_Point(_P(1),_P(2),_P(3))
41    ^CGA2_Point(_P(1),_P(2),_P(3))
42  }
43
44  DCGA_Point = { createPoint(_P(1),_P(2),_P(3)) }
45  Normalize_CGA1_Point = { _P(1)/(-_P(1).ei1) }
46  Normalize_CGA2_Point = { _P(1)/(-_P(1).ei2) }
47  Normalize_DCGA_Point = { _P(1)/(-_P(1).ei()) }
48  // Embed vector as CGA1 point
49  // (shorter name for CGA1_Point)
50  EV1 = { CGA1_Point(_P(1),_P(2),_P(3)) }
51  // Embed vector as CGA2 point
52  // (shorter name for CGA2_Point)
53  EV2 = { CGA2_Point(_P(1),_P(2),_P(3)) }
54  // Embed vector as DCGA point
55  // (shorter name for DCGA_Point)
56  EV = { DCGA_Point(_P(1),_P(2),_P(3)) }
57  // Project CGA1 point back to a vector
58  // in Euclidean 1 space
59  PV1 = { -(Normalize_CGA1_Point(_P(1)).IE1()).IE1() }
60  // Project CGA2 point back to a vector
61  //  in Euclidean _2_ space
62  PV2 = { -(Normalize_CGA2_Point(_P(1)).IE2()).IE2() }
63  // Project DCGA point back to a vector
64  // in Euclidean 1 space
65  PV = { PV1(_P(1).ei2) }
```

At first the origin and infinity of DCGA are defined as the outer products of the corresponding basis vectors of the both CGA copies. Then the Euclidean, conformal and double conformal pseudo scalars, dualization macros and point definitions are shown.

11.2 THE DCGA OBJECTS

The geometric objects of DCGA are defined in Chapt. 3 and Appendix A of [18].

11.2.1 Ellipsoid, Toroid and Sphere

Listing 11.3 shows how ellipsoid, toroid and sphere are defined in macros.clu.

Listing 11.3: Macros.clu for DCGA: ellipsoid, toroid and sphere.

```
1   // DCGA macros.clu
2
3   // DCGA point value-extraction operators
4   // for defining DCGA GIPNS 2-vector entities
5   Tx = { (1/2)*(e1^ei2+ei1^e6) }
6   Ty = { (1/2)*(e2^ei2+ei1^e7) }
7   Tz = { (1/2)*(e3^ei2+ei1^e8) }
8   Txy  = { (1/2)*(e7^e1+e6^e2) }
9   Tyz  = { (1/2)*(e7^e3+e8^e2) }
10  Tzx  = { (1/2)*(e8^e1+e6^e3) }
```

```
11  Txx = { e6^e1 }
12  Tyy = { e7^e2 }
13  Tzz = { e8^e3 }
14  Txt2 = { e1^eo2+eo1^e6 }
15  Tyt2 = { e2^eo2+eo1^e7 }
16  Tzt2 = { e3^eo2+eo1^e8 }
17  T1 = { -ei() }
18  Tt2 = { eo2^ei1+ei2^eo1 }
19  Tt4 = { -4*eo() }
20
21  // Ellipsoid(px,py,pz,rx,ry,rz)
22  // with center (px,py,pz), radii rx ry rz
23  Ellipsoid = {
24   pxSq = _P(1)*_P(1); pySq = _P(2)*_P(2);
25   pzSq = _P(3)*_P(3);
26   rxSq = _P(4)*_P(4); rySq = _P(5)*_P(5);
27   rzSq = _P(6)*_P(6);
28   -2*_P(1)*Tx()/rxSq + -2*_P(2)*Ty()/rySq
29   + -2*_P(3)*Tz()/rzSq +
30   Txx()/rxSq + Tyy()/rySq + Tzz()/rzSq +
31   (pxSq/rxSq + pySq/rySq + pzSq/rzSq - 1)*T1()
32  }
33  // Toroid(R,r) with major radius R and minor radius r
34  // Generalizes the Sphere(0,0,0,r) ~ Toroid(0,r)
35  Toroid = {
36   R = _P(1); r = _P(2); dSq = R*R - r*r;
37   Tt4() + 2*Tt2()*dSq + T1()*dSq*dSq - 4*R*R*(Txx()+Tyy())
38  }
39  // Sphere(x,y,z,r) with center point (x,y,z) and radius r
40  CGA1_Sphere = { CGA1_Point(_P(1),_P(2),_P(3))
41      - (1/2)*_P(4)*_P(4)*ei1 }
42  CGA2_Sphere = { CGA2_Point(_P(1),_P(2),_P(3))
43      - (1/2)*_P(4)*_P(4)*ei2 }
44  Sphere = {
45   CGA1_Sphere(_P(1),_P(2),_P(3),_P(4))
46   ^CGA2_Sphere(_P(1),_P(2),_P(3),_P(4))
47  }
```

First of all, some extraction operators have to be defined according to Table 1 of [18]. They are needed for many bivector entities of DCGA. Ellipsoid(px,py,pz,rx,ry,rz) is described based on 6 parameters with the center point (px,py,pz) and the three radii rx ry rz for each of the coordinate directions. Toroid(R,r) is defined based on its major radius R and the minor radius r. Sphere(x,y,z,r) is defined with centre point (x,y,z) and radius r as the outer product of spheres defined in each of the CGA copies. The following Listing 11.4 computes a sphere, an ellipsoid and a toroid and defines these objects to be visualized in red, magenta and blue.

Listing 11.4: GAALOPScript for the visualization of ellipsoid, toroid and sphere.

```
1  S = Sphere(1,2,3,2);          /* (px,py,pz,radius) */
2  E = Ellipsoid(1,2,3,4,3,2);   /* (px,py,pz,rx,ry,rz) */
3  T = Toroid(4,2);              /* radii R and r */
4
5  :Red;
```

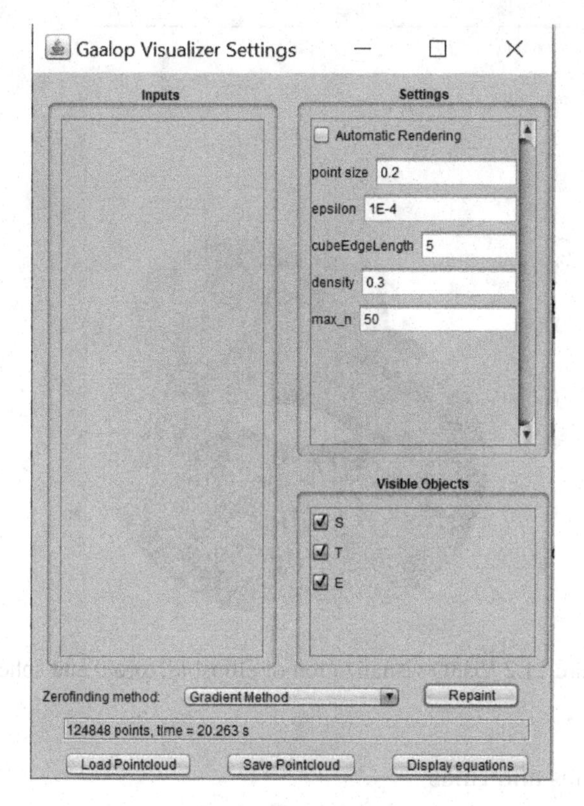

Figure 11.1 Vis3D settings.

```
6  :S;
7  :Magenta;
8  :E;
9  :Blue;
10 :T;
11 }
```

Before visualizing, Vis3D requests some settings according to Fig. 11.1. We change the density from the standard value 1 to 0.3 (which takes a bit longer for rendering).

After pushing the repaint button, the visualization is generated and in the settings panel the three objects S, T and E are indicated as visible objects. You are now able to control which object should be really visualized. Please find details about Vis3D in the master thesis [63]. Figure 11.2 shows the Vis3D visualization of ellipsoid, toroid and sphere according to Listing 11.4.

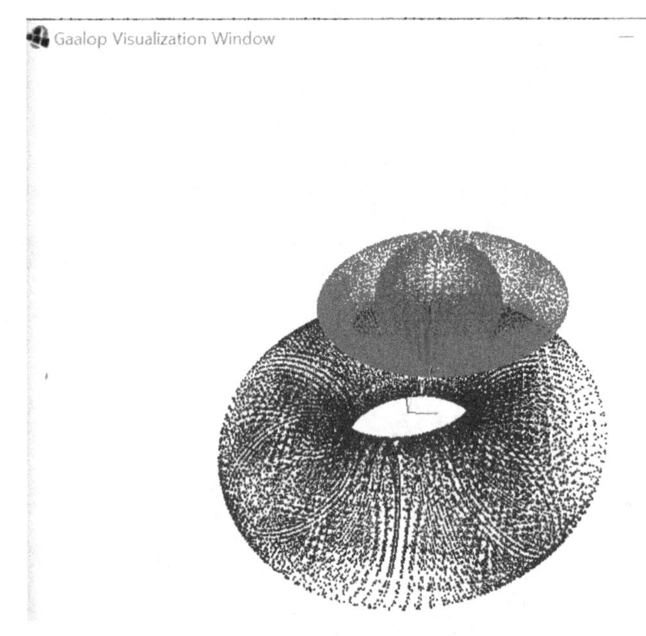

Figure 11.2 Vis3D visualization of ellipsoid, toroid and sphere.

11.2.2 Planes and Lines

Listing 11.5 shows the DCGA definitions of planes and lines based on the outer products of their definitions in each of the two CGA copies.

Listing 11.5: Macros.clu for DCGA: planes and lines.

```
1  // DCGA macros.clu
2
3  // Plane(nx,ny,nz,d) with normal (nx,ny,nz)
4  // at distance d from origin
5  CGA1_Plane = {
6    Normalize(_P(1)*e1 + _P(2)*e2 + _P(3)*e3)
7    + _P(4)*ei1
8  }
9  CGA2_Plane = {
10   Normalize(_P(1)*e6 + _P(2)*e7 + _P(3)*e8)
11   + _P(4)*ei2
12 }
13 Plane = {
14  CGA1_Plane(_P(1),_P(2),_P(3),_P(4))
15 ^CGA2_Plane(_P(1),_P(2),_P(3),_P(4))
16 }
17 // Line(px,py,pz,dx,dy,dz) through (px,py,pz)
```

```
18 // in direction of (dx,dy,dz)
19 CGA1_Line = {
20   d = Normalize(_P(4)*e1 + _P(5)*e2 + _P(6)*e3);
21   D = E1D(d);
22   p = _P(1)*e1 + _P(2)*e2 + _P(3)*e3;
23   D - (p.D)*ei1
24 }
25 CGA2_Line = {
26   d = Normalize(_P(4)*e6 + _P(5)*e7 + _P(6)*e8);
27   D = E2D(d);
28   p = _P(1)*e6 + _P(2)*e7 + _P(3)*e8;
29   D - (p.D)*ei2
30 }
31 Line = {
32   CGA1_Line(_P(1),_P(2),_P(3),_P(4),_P(5),_P(6))
33  ^CGA2_Line(_P(1),_P(2),_P(3),_P(4),_P(5),_P(6))
34 }
```

Figure 11.3 shows the Vis3D visualization of plane and line according to Listing 11.6.

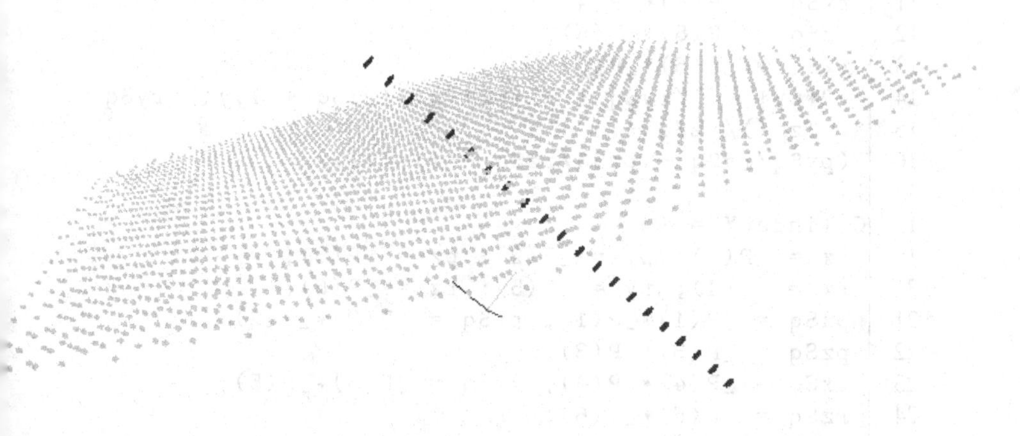

Figure 11.3 Vis3D visualization of plane and line.

Listing 11.6: GAALOPScript for the visualization of plane and line.

```
1 P = Plane(1,2,3,sqrt(1+4+9)); /* (nx,ny,nz,distance) */
2 L = Line(1,2,3,0,0,1);        /* (px,py,pz,dx,dy,dz) */
3
4 :Green;
```

```
5  :P;
6  :Blue;
7  :L;
8  }
```

11.2.3 Cylinders

Listing 11.7 shows the DCGA definitions of cylinders aligned in the direction of the coordinate axis.

Listing 11.7: Macros.clu for DCGA: cylinders.

```
1  // DCGA macros.clu
2
3  // Cylinder(px,py,pz,rx,ry,rz)
4  // with center (px,py,pz) and radii rx ry rz
5  CylinderX = {
6   px = _P(1); py = _P(2); pz = _P(3);
7   rx = _P(4); ry = _P(5); rz = _P(6);
8   pxSq = _P(1)*_P(1);
9   pySq = _P(2)*_P(2);
10  pzSq = _P(3)*_P(3);
11  rxSq = _P(4)*_P(4);
12  rySq = _P(5)*_P(5);
13  rzSq = _P(6)*_P(6);
14  -2*py*Ty()/rySq + -2*pz*Tz()/rzSq + Tyy()/rySq
15  + Tzz()/rzSq +
16  (pySq/rySq + pzSq/rzSq - 1)*T1()
17 }
18 CylinderY = {
19  px = _P(1); py = _P(2); pz = _P(3);
20  rx = _P(4); ry = _P(5); rz = _P(6);
21  pxSq = _P(1)*_P(1); pySq = _P(2)*_P(2);
22  pzSq = _P(3)*_P(3);
23  rxSq = _P(4)*_P(4); rySq = _P(5)*_P(5);
24  rzSq = _P(6)*_P(6);
25  -2*px*Tx()/rxSq + -2*pz*Tz()/rzSq + Txx()/rxSq
26  + Tzz()/rzSq +
27  (pxSq/rxSq + pzSq/rzSq - 1)*T1()
28 }
29 CylinderZ = {
30  px = _P(1); py = _P(2); pz = _P(3);
31  rx = _P(4); ry = _P(5); rz = _P(6);
32  pxSq = _P(1)*_P(1); pySq = _P(2)*_P(2);
33  pzSq = _P(3)*_P(3);
34  rxSq = _P(4)*_P(4); rySq = _P(5)*_P(5);
```

```
35  rzSq = _P(6)*_P(6);
36  -2*px*Tx()/rxSq + -2*py*Ty()/rySq + Txx()/rxSq
37  + Tyy()/rySq +
38  (pxSq/rxSq + pySq/rySq - 1)*T1()
39 }
```

Figure 11.4 shows the Vis3D visualization of cylinders according to Listing 11.8.

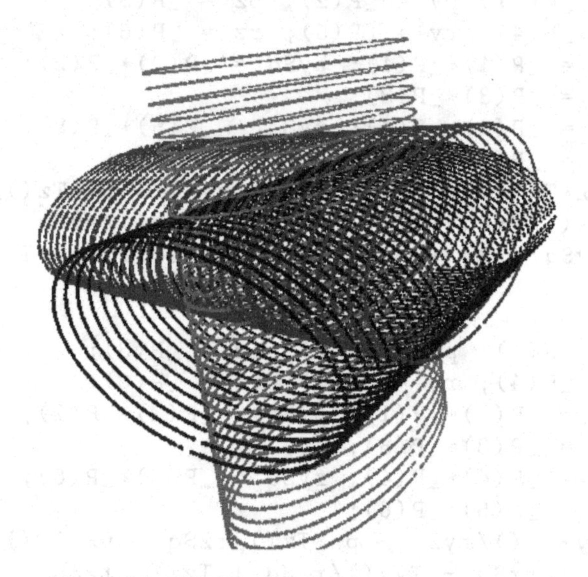

Figure 11.4 Vis3D visualization of cylinders.

Listing 11.8: GAALOPScript for the visualization of cylinders.

```
1  CylX = CylinderX(0,1,0,1,2,4);
2  CylY = CylinderY(1,0,0,2,1,4);
3  CylZ = CylinderZ(1,0,0,4,2,1);
4
5  :Blue;
6  :CylX;
7  :Red;
8  :CylY;
9  :Black;
10 :CylZ;
```

11.2.4 Cones

Listing 11.9 shows the DCGA definitions of cones.

Listing 11.9: Macros.clu for DCGA: cones.

```
1  // DCGA macros.clu
2
3  // Cone(px,py,pz,rx,ry,rz) with center (px,py,pz)
4  // and radii rx ry rz
5  ConeX = {
6   px = _P(1); py = _P(2); pz = _P(3);
7   rx = _P(4); ry = _P(5); rz = _P(6);
8   pxSq = _P(1)*_P(1); pySq = _P(2)*_P(2);
9   pzSq = _P(3)*_P(3);
10  rxSq = _P(4)*_P(4); rySq = _P(5)*_P(5);
11  rzSq = _P(6)*_P(6);
12  2*(px*Tx()/rxSq - py*Ty()/rySq - pz*Tz()/rzSq)
13  - Txx()/rxSq + Tyy()/rySq + Tzz()/rzSq
14  + (pySq/rySq + pzSq/rzSq - pxSq/rxSq)*T1()
15 }
16 ConeY = {
17  px = _P(1); py = _P(2); pz = _P(3);
18  rx = _P(4); ry = _P(5); rz = _P(6);
19  pxSq = _P(1)*_P(1); pySq = _P(2)*_P(2);
20  pzSq = _P(3)*_P(3);
21  rxSq = _P(4)*_P(4); rySq = _P(5)*_P(5);
22  rzSq = _P(6)*_P(6);
23  2*(py*Ty()/rySq - pz*Tz()/rzSq - px*Tx()/rxSq)
24  +Txx()/rxSq - Tyy()/rySq + Tzz()/rzSq
25  + (pxSq/rxSq - pySq/rySq + pzSq/rzSq)*T1()
26 }
27 ConeZ = {
28  px = _P(1); py = _P(2); pz = _P(3);
29  rx = _P(4); ry = _P(5); rz = _P(6);
30  pxSq = _P(1)*_P(1); pySq = _P(2)*_P(2);
31  pzSq = _P(3)*_P(3);
32  rxSq = _P(4)*_P(4); rySq = _P(5)*_P(5);
33  rzSq = _P(6)*_P(6);
34  2*(pz*Tz()/rzSq - py*Ty()/rySq - px*Tx()/rxSq)
35  + Txx()/rxSq + Tyy()/rySq - Tzz()/rzSq
36  + (pxSq/rxSq + pySq/rySq - pzSq/rzSq)*T1()
37 }
```

Figure 11.5 shows the Vis3D visualization of the cones CneY and CneZ according to Listing 11.10.

Listing 11.10: GAALOPScript for the visualization of cones.

```
1  // Cones   (px,py,pz,rx,ry,rz)
2  CneX = ConeX(1,0,0,1,1,1);
3  CneY = ConeY(0,1,0,1,1,1);
```

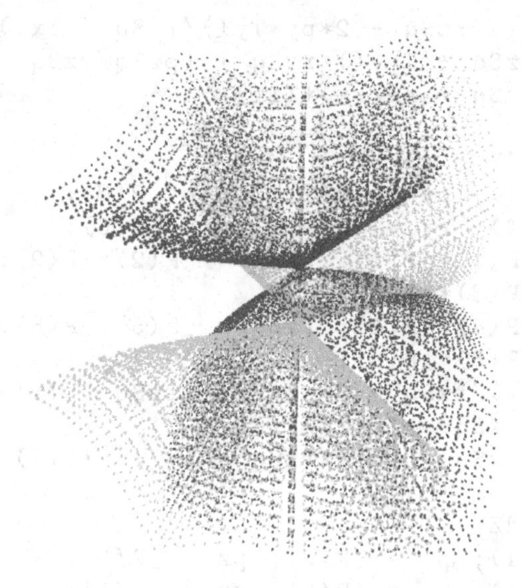

Figure 11.5 Vis3D visualization of cones.

```
4  CneZ = ConeZ(0,0,1,1,1,1);
5
6  :Black;
7  :CneX;
8  :Red;
9  :CneY;
10 :Green;
11 :CneZ;
```

11.2.5 Paraboloids

Listing 11.11 shows the DCGA definitions of paraboloids.

Listing 11.11: Macros.clu for DCGA: paraboloids.

```
1  // DCGA macros.clu
2
3  // Paraboloid(px,py,pz,rx,ry,rz)
4  // with vertex (px,py,pz) and radii rx ry rz
5  ParaboloidX = {
6   px = _P(1); py = _P(2); pz = _P(3);
7   rx = _P(4); ry = _P(5); rz = _P(6);
8   pxSq = _P(1)*_P(1); pySq = _P(2)*_P(2);
9   pzSq = _P(3)*_P(3);
10  rxSq = _P(4)*_P(4); rySq = _P(5)*_P(5);
11  rzSq = _P(6)*_P(6);
```

```
12   -2*pz*Tz()/rzSq - 2*py*Ty()/rySq - Tx()/rx
13   +Tzz()/rzSq + Tyy()/rySq + (pzSq/rzSq
14   + pySq/rySq + px/rx)*T1()
15  }
16  ParaboloidY = {
17   px = _P(1); py = _P(2); pz = _P(3);
18   rx = _P(4); ry = _P(5); rz = _P(6);
19   pxSq = _P(1)*_P(1); pySq = _P(2)*_P(2);
20   pzSq = _P(3)*_P(3);
21   rxSq = _P(4)*_P(4); rySq = _P(5)*_P(5);
22   rzSq = _P(6)*_P(6);
23   -2*px*Tx()/rxSq - 2*pz*Tz()/rzSq - Ty()/ry
24   + Txx()/rxSq + Tzz()/rzSq
25   + (pxSq/rxSq + pzSq/rzSq + py/ry)*T1()
26  }
27  ParaboloidZ = {
28   px = _P(1); py = _P(2); pz = _P(3);
29   rx = _P(4); ry = _P(5); rz = _P(6);
30   pxSq = _P(1)*_P(1); pySq = _P(2)*_P(2);
31   pzSq = _P(3)*_P(3);
32   rxSq = _P(4)*_P(4); rySq = _P(5)*_P(5);
33   rzSq = _P(6)*_P(6);
34   -2*px*Tx()/rxSq - 2*py*Ty()/rySq - Tz()/rz
35   + Txx()/rxSq + Tyy()/rySq
36   + (pxSq/rxSq + pySq/rySq + pz/rz)*T1()
37  }
38  // Hyperbolic paraboloid (px,py,pz,rx,ry,rz)
39  // z-axis aligned
40  // with center point (px,py,pz) and radii rx ry rz
41  HParaboloidZ = {
42   px = _P(1); py = _P(2); pz = _P(3);
43   rx = _P(4); ry = _P(5); rz = _P(6);
44   pxSq = _P(1)*_P(1); pySq = _P(2)*_P(2);
45   pzSq = _P(3)*_P(3);
46   rxSq = _P(4)*_P(4); rySq = _P(5)*_P(5);
47   rzSq = _P(6)*_P(6);
48   -2*px*Tx()/rxSq + 2*py*Ty()/rySq - Tz()/rz
49   + Txx()/rxSq - Tyy()/rySq
50   + (pxSq/rxSq - pySq/rySq + pz/rz)*T1()
51  }
```

Figure 11.6 shows the Vis3D visualization of paraboloids according to Listing 11.12.

Listing 11.12: GAALOPScript for the visualization of paraboloids.

```
1  // Paraboloids (px,py,pz,rx,ry,rz)
```

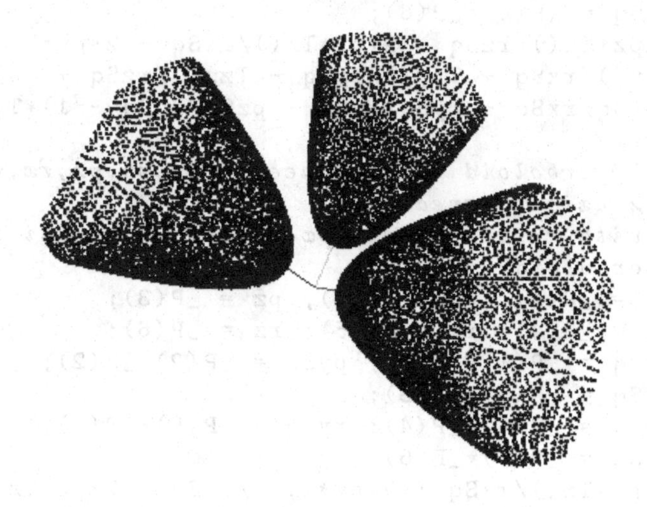

Figure 11.6 Vis3D visualization of paraboloids.

```
2  ParX = ParaboloidX(1,0,0,1,1,1);
3  ParY = ParaboloidY(0,1,0,1,1,1);
4  ParZ = ParaboloidZ(0,0,1,1,1,1);
5
6  :Black;
7  :ParX;
8  :ParY;
9  :ParZ;
```

11.2.6 Hyperboloids

Listing 11.13 shows the DCGA definitions of hyperboloids.

Listing 11.13: Macros.clu for DCGA: hyperboloids.

```
1  // DCGA macros.clu
2
3  // Hyperboloid of one sheet (px,py,pz,rx,ry,rz)
4  // z-axis aligned
5  // with center point (px,py,pz) and radii rx ry rz
6  Hyperboloid1 = {
7   px = _P(1); py = _P(2); pz = _P(3);
8   rx = _P(4); ry = _P(5); rz = _P(6);
9   pxSq = _P(1)*_P(1); pySq = _P(2)*_P(2);
10  pzSq = _P(3)*_P(3);
11  rxSq = _P(4)*_P(4); rySq = _P(5)*_P(5);
```

```
12   rzSq = _P(6)*_P(6);
13   2*pz*Tz()/rzSq - 2*px*Tx()/rxSq - 2*py*Ty()/rySq +
14   Txx()/rxSq + Tyy()/rySq - Tzz()/rzSq +
15   (pxSq/rxSq + pySq/rySq - pzSq/rzSq - 1)*T1()
16 }
17 // Hyperboloid of two sheets (px,py,pz,rx,ry,rz)
18 // z-axis aligned
19 // with center point (px,py,pz) and radii rx ry rz
20 Hyperboloid2 = {
21   px = _P(1); py = _P(2); pz = _P(3);
22   rx = _P(4); ry = _P(5); rz = _P(6);
23   pxSq = _P(1)*_P(1); pySq = _P(2)*_P(2);
24   pzSq = _P(3)*_P(3);
25   rxSq = _P(4)*_P(4); rySq = _P(5)*_P(5);
26   rzSq = _P(6)*_P(6);
27   2*px*Tx()/rxSq + 2*py*Ty()/rySq - 2*pz*Tz()/rzSq -
28   Txx()/rxSq - Tyy()/rySq + Tzz()/rzSq +
29   (pzSq/rzSq - pxSq/rxSq - pySq/rySq - 1)*T1()
30 }
```

Figure 11.7 shows the Vis3D visualization of the hyperboloid Hyp1 according to Listing 11.14.

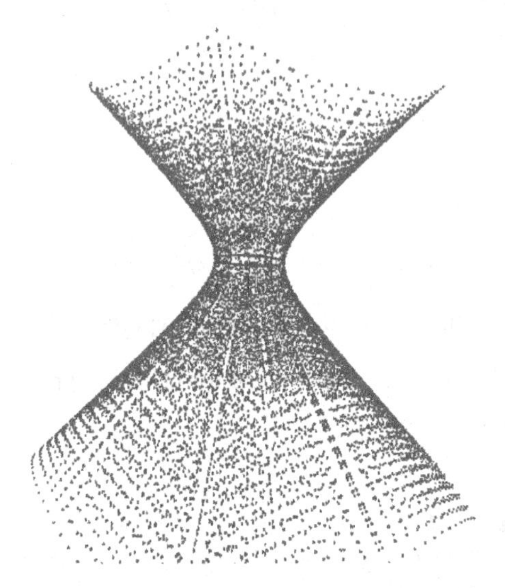

Figure 11.7 Vis3D visualization of an hyperboloid.

Listing 11.14: GAALOPScript for the visualization of hyperboloids.

```
1 // Hyperbolic paraboloid (px,py,pz,rx,ry,rz)
```

```
2  // z-axis aligned
3  HparZ = HParaboloidZ(0,0,1,1,1,1);
4  // Hyperboloid of one sheet (px,py,pz,rx,ry,rz)
5  // z-axis aligned
6  Hyp1 = Hyperboloid1(0,0,1,1,1,1);
7  // Hyperboloid of two sheets (px,py,pz,rx,ry,rz)
8  // z-axis aligned
9  Hyp2 = Hyperboloid2(0,0,1,1,1,1);
10
11 :Black;
12 :HparZ;
13 :Red;
14 :Hyp1;
15 :Green;
16 :Hyp2;
```

11.2.7 Parabolic and Hyperbolic Cylinders

Listing 11.15 shows the DCGA definitions of parabolic and hyperbolic cylinders.

Listing 11.15: Macros.clu for DCGA: parabolic and hyperbolic cylinders.

```
1  // DCGA macros.clu
2
3  // Parabolic cylinders (px,py,pz,rx,ry,rz)
4  // with center point (px,py,pz) and radii rx ry rz
5  PCylinderX = {
6   px = _P(1); py = _P(2); pz = _P(3);
7   rx = _P(4); ry = _P(5); rz = _P(6);
8   pxSq = _P(1)*_P(1); pySq = _P(2)*_P(2);
9   pzSq = _P(3)*_P(3);
10  rxSq = _P(4)*_P(4); rySq = _P(5)*_P(5);
11  rzSq = _P(6)*_P(6);
12  -2*py*Ty()/rySq - Tz()/rz + Tyy()/rySq
13  +(pySq/rySq + pz/rz)*T1()
14 }
15 PCylinderY = {
16  px = _P(1); py = _P(2); pz = _P(3);
17  rx = _P(4); ry = _P(5); rz = _P(6);
18  pxSq = _P(1)*_P(1); pySq = _P(2)*_P(2);
19  pzSq = _P(3)*_P(3);
20  rxSq = _P(4)*_P(4); rySq = _P(5)*_P(5);
21  rzSq = _P(6)*_P(6);
22  -2*px*Tx()/rxSq - Tz()/rz
23  +Txx()/rxSq + (pxSq/rxSq + pz/rz)*T1()
24 }
25 PCylinderZ = {
```

```
26    px = _P(1); py = _P(2); pz = _P(3);
27    rx = _P(4); ry = _P(5); rz = _P(6);
28    pxSq = _P(1)*_P(1); pySq = _P(2)*_P(2);
29    pzSq = _P(3)*_P(3);
30    rxSq = _P(4)*_P(4); rySq = _P(5)*_P(5);
31    rzSq = _P(6)*_P(6);
32    -2*px*Tx()/rxSq - Ty()/ry
33    + Txx()/rxSq + (pxSq/rxSq + py/ry)*T1()
34   }
35
36   // Hyperbolic cylinders (px,py,pz,rx,ry,rz)
37   // with center point (px,py,pz) and radii rx ry rz
38   HCylinderX = {
39    px = _P(1); py = _P(2); pz = _P(3);
40    rx = _P(4); ry = _P(5); rz = _P(6);
41    pxSq = _P(1)*_P(1); pySq = _P(2)*_P(2);
42    pzSq = _P(3)*_P(3);
43    rxSq = _P(4)*_P(4); rySq = _P(5)*_P(5);
44    rzSq = _P(6)*_P(6);
45    -2*py*Ty()/rySq + 2*pz*Tz()/rzSq + Tyy()/rySq
46    - Tzz()/rzSq +
47    (pySq/rySq - pzSq/rzSq - 1)*T1()
48   }
49   HCylinderY = {
50    px = _P(1); py = _P(2); pz = _P(3);
51    rx = _P(4); ry = _P(5); rz = _P(6);
52    pxSq = _P(1)*_P(1); pySq = _P(2)*_P(2);
53    pzSq = _P(3)*_P(3);
54    rxSq = _P(4)*_P(4); rySq = _P(5)*_P(5);
55    rzSq = _P(6)*_P(6);
56    2*px*Tx()/rxSq - 2*pz*Tz()/rzSq - Txx()/rxSq
57    + Tzz()/rzSq +
58    (-pxSq/rxSq + pzSq/rzSq - 1)*T1()
59   }
60   HCylinderZ = {
61    px = _P(1); py = _P(2); pz = _P(3);
62    rx = _P(4); ry = _P(5); rz = _P(6);
63    pxSq = _P(1)*_P(1); pySq = _P(2)*_P(2);
64    pzSq = _P(3)*_P(3);
65    rxSq = _P(4)*_P(4); rySq = _P(5)*_P(5);
66    rzSq = _P(6)*_P(6);
67    -2*px*Tx()/rxSq + 2*py*Ty()/rySq + Txx()/rxSq
68    - Tyy()/rySq +
69    (pxSq/rxSq - pySq/rySq - 1)*T1()
70   }
```

11.2.8 Specific Planes

Listing 11.16 shows the DCGA definitions of specific planes.

Listing 11.16: Macros.clu for DCGA: specific planes.

```

// Parallel planes pair (p1,p2) with planes a=p1, a=p2
// perpendicular to a-axis
PPlanesX = {
 p1 = _P(1); p2 = _P(2);
 Txx() - (p1+p2)*Tx() + p1*p2*T1()
}
PPlanesY = {
 p1 = _P(1); p2 = _P(2);
 Tyy() - (p1+p2)*Ty() + p1*p2*T1()
}
PPlanesZ = {
 p1 = _P(1); p2 = _P(2);
 Tzz() - (p1+p2)*Tz() + p1*p2*T1()
}
// Non-parallel planes pairs (also types of cylinders)
// XPlanesX(y,z,ry,rz) x-axis aligned,
// with vertex (y,z) and slope (+|-)rz/ry
XPlanesX = {
 py = _P(1); pz = _P(2); ry = _P(3); rz = _P(4);
 pySq = _P(1)*_P(1); pzSq = _P(2)*_P(2);
 rySq = _P(3)*_P(3); rzSq = _P(4)*_P(4);
 -2*py*Ty()/rySq + 2*pz*Tz()/rzSq + Tyy()/rySq
 - Tzz()/rzSq + (pySq/rySq - pzSq/rzSq)*T1()
}
// XPlanesY(x,z,rx,rz) y-axis aligned,
// with vertex (x,z) and slope (+|-)rz/rx
XPlanesY = {
 px = _P(1); pz = _P(2); rx = _P(3); rz = _P(4);
 pxSq = _P(1)*_P(1); pzSq = _P(2)*_P(2);
 rxSq = _P(3)*_P(3); rzSq = _P(4)*_P(4);
 -2*pz*Tz()/rzSq + 2*px*Tx()/rxSq + Tzz()/rzSq
 - Txx()/rxSq + (pzSq/rzSq - pxSq/rxSq)*T1()
}
// XPlanesZ(x,y,rx,ry) z-axis aligned,
// with vertex (x,y) and slope (+|-)ry/rx
XPlanesZ = {
 px = _P(1); py = _P(2); rx = _P(3); ry = _P(4);
 pxSq = _P(1)*_P(1); pySq = _P(2)*_P(2);
 rxSq = _P(3)*_P(3); rySq = _P(4)*_P(4);
 -2*px*Tx()/rxSq + 2*py*Ty()/rySq + Txx()/rxSq
 - Tyy()/rySq + (pxSq/rxSq - pySq/rySq)*T1()
}
```

11.2.9 Cyclides

Listing 11.17 shows the DCGA definitions of cyclides.

Listing 11.17: Macros.clu for DCGA: cyclides.

```
1
2  // DupinCyclide(R,r1,r2) with major radius R
3  // and minor radii r1 and r2
4  // Generalizes the Toroid(R,r) ~ DupinCyclide(R,r,r)
5  DupinCyclide = {
6   u = (1/2)*(_P(2) + _P(3)); c = (1/2)*(_P(2) - _P(3));
7   a = _P(1); bSq = a*a-c*c;
8   Tt4() + 2*Tt2()*(bSq-u*u) - 4*a*a*Txx() - 4*bSq*Tyy() +
9   8*a*c*u*Tx() + ((bSq-u*u)*(bSq-u*u) - 4*c*c*u*u)*T1()
10 }
11 hornedDupinCyclide = {
12  u = (1/2)*(_P(2) + _P(3)); c = (1/2)*(_P(2) - _P(3));
13  a = _P(1); bSq = a*a-u*u;
14  Tt4() + 2*Tt2()*(bSq-c*c) - 4*a*a*Txx() - 4*bSq*Tyy() +
15  8*a*c*u*Tx() + ((bSq-c*c)*(bSq-c*c) - 4*c*c*u*u)*T1()
16 }
```

Figure 11.8 shows the Vis3D visualization of cyclides according to Listing 11.18.

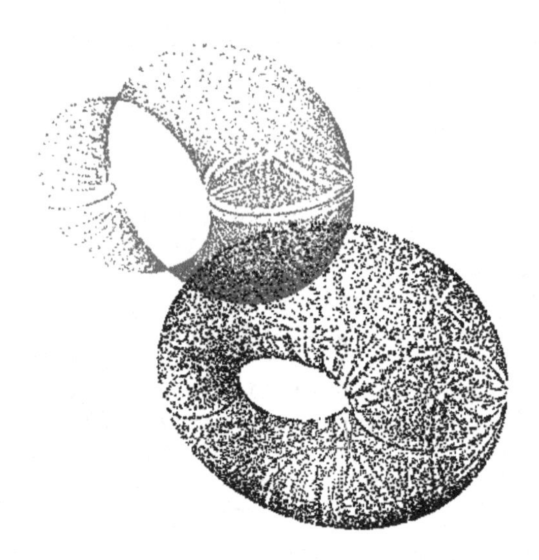

Figure 11.8 Vis3D visualization of cyclides.

Listing 11.18: GAALOPScript for the visualization of cyclides.

```
1  // DupinCyclide and hornedDupinCyclide(R,r1,r1)
2  // also translated
```

```
3  :Black;
4  :DC = Translator(1,1,1) * DupinCyclide(3,2,1)
5      * ~Translator(1,1,1);
6  :Red;
7  :HDC = Translator(5,5,5) * hornedDupinCyclide(3,2,1)
8      * ~Translator(5,5,5);
```

11.3 THE DCGA TRANSFORMATIONS

Listing 11.19 shows the predefined functionality for DCGA transformations.

Listing 11.19: Macros.clu for DCGA: transformations.

```
1  // DCGA macros.clu
2
3  // Convert degrees angle to radians
4  Deg2Rad = { _P(1)*acos(-1)/180 }
5
6  // Rotor(x,y,z,a) with axis (x,y,z)
7  // and rotation angle a in _degrees_
8  CGA1_Rotor = {
9   t = Deg2Rad(_P(4));
10  cos(t/2) + sin(t/2)*E1D(Normalize(_P(1)*e1
11    + _P(2)*e2 + _P(3)*e3))
12 }
13 CGA2_Rotor = { t = Deg2Rad(_P(4));
14  cos(t/2) + sin(t/2)*E2D(Normalize(_P(1)*e6
15   + _P(2)*e7 + _P(3)*e8))
16 }
17 Rotor = {
18  CGA1_Rotor(_P(1),_P(2),_P(3),_P(4))
19 ^CGA2_Rotor(_P(1),_P(2),_P(3),_P(4))
20 }
21 // Dilator(d) with scalar dilation factor d
22 CGA1_Dilator = {
23  (1/2)*(1+_P(1)) + (1/2)*(1-_P(1))*(ei1^eo1)
24 }
25 CGA2_Dilator = {
26  (1/2)*(1+_P(1)) + (1/2)*(1-_P(1))*(ei2^eo2)
27 }
28 Dilator = {
29  CGA1_Dilator(_P(1))^CGA2_Dilator(_P(1))
30 }
31 // Translator(x,y,z) for translation
32 // by displacement vector (x,y,z)
33 CGA1_Translator = {
```

```
34    1 - (1/2)*(_P(1)*e1+_P(2)*e2+_P(3)*e3)*ei1
35  }
36  CGA2_Translator = {
37    1 - (1/2)*(_P(1)*e6+_P(2)*e7+_P(3)*e8)*ei2
38  }
39  Translator = {
40    CGA1_Translator(_P(1),_P(2),_P(3))
41  ^CGA2_Translator(_P(1),_P(2),_P(3))
42  }
```

Rotations, dilations and translations are defined as outer products of the corresponding transformations in the two CGA copies.

Listing 11.20 computes transformations of an ellipsoid and visualizes them.

Listing 11.20: GAALOPScript for transformations of an ellipsoid.

```
1   // Rotate, dilate, and translate an ellipsoid
2   ER = Rotor(0,0,1,45) * Ellipsoid(0,0,0,1,3,1)
3      * ~Rotor(0,0,1,45);
4   ED = Dilator(2) * Ellipsoid(0,0,0,1,2,3) * ~Dilator(2);
5   ET = Translator(1,2,3) * Ellipsoid(0,0,0,2,3,4)
6      * ~Translator(1,2,3);
7
8   :Red;
9   :ER;
10  :Green;
11  :ED;
12  :Blue;
13  :ET;
```

Figure 11.9 shows the Vis3D visualization of a rotation, a dilation and a translation of an ellipsoid according to Listing 11.20.

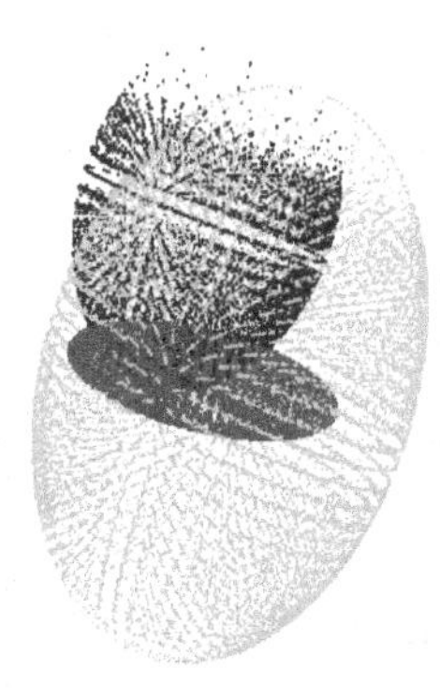

Figure 11.9 Vis3D visualization of transformations of an ellipsoid.

11.4 INTERSECTIONS

Intersections can be computed as outer products of geometric objects. Listing 11.21, for instance, computes the intersection of an ellipsoid and a line and visualizes it.

Listing 11.21: GAALOPScript for the intersection of ellipsoid and line.

```
1  // Ellipsoid-Line intersection test
2
3  E = Ellipsoid(0,0,0,2,3,4);
4  L = Line(0,0,0,1,1,1);
5  EL = E^L;
6
7  :Cyan;
8  :E;
9  :Yellow;
10 :L;
11 :Red;
12 :EL;
```

Figure 11.10 shows the Vis3D visualization. You can see the intersecting point pair if you uncheck the line L in the Visible Objects part of the Visualizer settings.

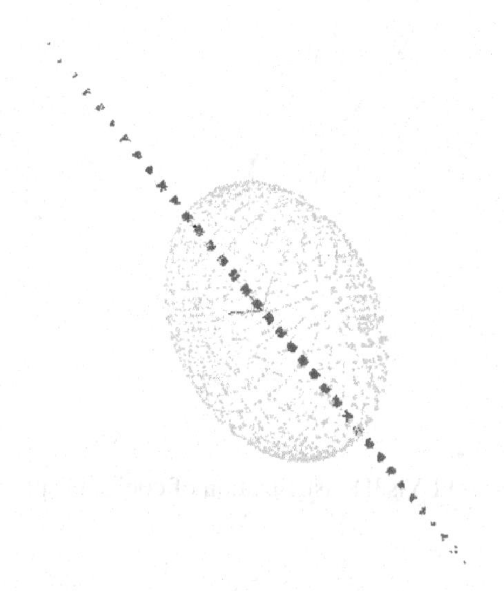

Figure 11.10 Vis3D visualization of the intersection of an ellipsoid and a line.

Intersections of quadrics and a plane result in conics. Listing 11.22 computes some conics in space and visualizes them.

The ellipse e, for instance, is computed as the intersection of an elliptic cylinder with the plane.

Listing 11.22: GAALOPScript for the visualization of conics in space.

```
1  // Ellipse, parabola, hyperbola
2  :e = CylinderZ(0,0,0,1,4,1)^Plane(0,0,1,1);
3  :p = PCylinderZ(0,0,0,1,4,1)^Plane(0,0,1,1);
4  :h = HCylinderZ(0,0,0,1,4,1)^Plane(0,0,1,1);
5
6  :Red;
7  :e;
8  :Blue;
9  :p;
10 :Green;
11 :h;
```

Figure 11.11 shows the Vis3D visualization of conics according to Listing 11.22.

The paper [19] shows some examples regarding conic and cyclidic sections with computing and visulization using GAALOP.

Figure 11.11 Vis3D visualization of conics in space.

11.5 REFLECTIONS AND PROJECTIONS

Listing 11.23, for instance, computes the reflection and projection of a line onto a sphere.

Listing 11.23: GAALOPScript for the reflection and projection of a line.

```
1  // Reflection and projection of line onto sphere
2
```

```
3  S = Sphere(0,0,0,3);
4  L = Line(0,0,4,1,0,0);
5  R = S*L*~S;
6  P = (L.S)*~S;
7
8  :Cyan;
9  :S;
10 :Magenta;
11 :L;
12 :Yellow;
13 :R;
14 :Red;
15 :P;
```

Figure 11.12 shows the Vis3D visualization.

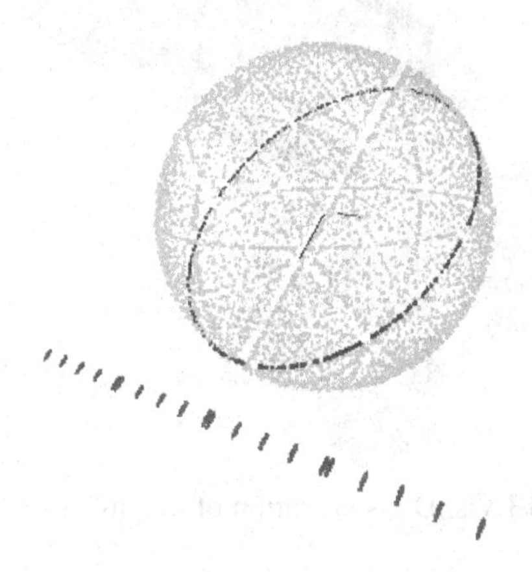

Figure 11.12 Vis3D visualization of the reflection and projection of a line onto a sphere.

11.6 INVERSIONS

Listing 11.24, for instance, computes the inversion of a toroid in a sphere.

Listing 11.24: GAALOPScript for the inversion of a toroid.

```
1  // Parabolic cyclide as inversion of Toroid in Sphere
2  // centered on toroid surface
3  T = Toroid(5,3);
4  PC = Sphere(2,0,0,3) * T * ~Sphere(2,0,0,3);
```

```
5  // PC rotated 60deg around the z-axis
6  RPC = Rotor(0,0,1,60) * PC * ~Rotor(0,0,1,60);
7
8  :Green;
9  :T;
10 :Red;
11 :PC;
12 :Blue;
13 :RPC;
```

Figure 11.13 shows the Vis3D visualization of the inversion of a toroid according to Listing 11.24.

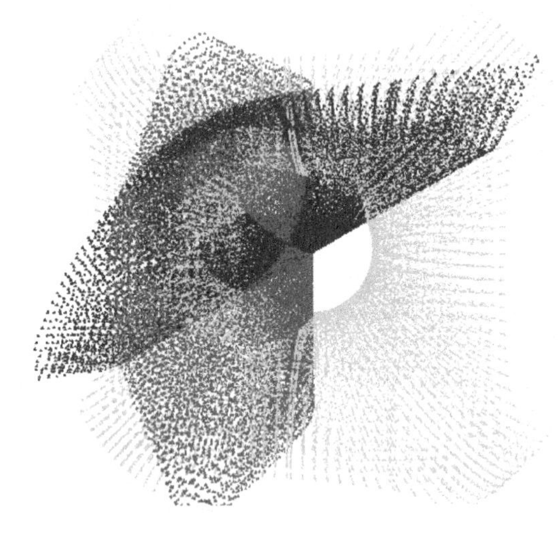

Figure 11.13 Vis3D visualization of the inversion of a toroid.

CHAPTER 12

Geometric Algebra for Cubics

CONTENTS

The cubic Conformal Geometric Algebra (CCGA) is based on the paper "Cubic Curves and Cubic Surfaces from Contact Points in Conformal Geometric Algebra" [38]. With CCGA, we can show that GAALOP is able to handle a Geometric Algebra with $2^{16} = 65.536$-dimensional multivectors.

12.1 GAALOP DEFINITION

Listing 12.1 presents the basic definitions for CCGA.

Listing 12.1: Definition.csv for CCGA.

```
1  1,e1,e2,eo1,ei1,eo2,ei2,eo3,ei3,eo4,ei4,eo5,ei5,eo6,ei6,eo7,ei7
2  ep1=0.7071067811865475*ei1-0.7071067811865475*eo1,
     ep2=0.7071067811865475*ei2-0.7071067811865475*eo2,
     ep3=0.7071067811865475*ei3-0.7071067811865475*eo3,
     ep4=0.7071067811865475*ei4-0.7071067811865475*eo4,
     ep5=0.7071067811865475*ei5-0.7071067811865475*eo5,
     ep6=0.7071067811865475*ei6-0.7071067811865475*eo6,
     ep7=0.7071067811865475*ei7-0.7071067811865475*eo7,
     em1=0.7071067811865475*ei1+0.7071067811865475*eo1,
     em2=0.7071067811865475*ei2+0.7071067811865475*eo2,
     em3=0.7071067811865475*ei3+0.7071067811865475*eo3,
     em4=0.7071067811865475*ei4+0.7071067811865475*eo4,
     em5=0.7071067811865475*ei5+0.7071067811865475*eo5,
     em6=0.7071067811865475*ei6+0.7071067811865475*eo6,
     em7=0.7071067811865475*ei7+0.7071067811865475*eo7
3  1,e1,e2,ep1,ep2,ep3,ep4,ep5,ep6,ep7,em1,em2,em3,em4,em5,em6,em7
4  e1=1,e2=1,ep1=1,ep2=1,ep3=1,ep4=1,ep5=1,ep6=1,ep7=1,
     em1=-1,em2=-1,em3=-1,em4=-1,em5=-1,em6=-1,em7=-1
5  eo1=0.7071067811865475*em1-0.7071067811865475*ep1,
     eo2=0.7071067811865475*em2-0.7071067811865475*ep2,
     eo3=0.7071067811865475*em3-0.7071067811865475*ep3,
```

DOI: 10.1201/9781003139003-12

```
eo4=0.7071067811865475*em4-0.7071067811865475*ep4,
eo5=0.7071067811865475*em5-0.7071067811865475*ep5,
eo6=0.7071067811865475*em6-0.7071067811865475*ep6,
eo7=0.7071067811865475*em7-0.7071067811865475*ep7,
ei1=0.7071067811865475*ep1+0.7071067811865475*em1,
ei2=0.7071067811865475*ep2+0.7071067811865475*em2,
ei3=0.7071067811865475*ep3+0.7071067811865475*em3,
ei4=0.7071067811865475*ep4+0.7071067811865475*em4,
ei5=0.7071067811865475*ep5+0.7071067811865475*em5,
ei6=0.7071067811865475*ep6+0.7071067811865475*em6,
ei7=0.7071067811865475*ep7+0.7071067811865475*em7
```

The structure of the CCGA consists of the two basis vectors of the 2D plane and additionally 7 basis elements squaring to 1 and 7 basis elements squaring to -1 which are transformed according to Eq. (1) of [38].

Line 3 defines the 16 standard basis vectors with their signatures in the lines 4. Line 1 shows the used basis vectors and the line 2 as well as the line 5 present all the transformations between the basis vectors. Listing 12.2 presents the basic functionality.

Listing 12.2: Macros.clu for CCGA.

```
1  // Cubic CGA macros
2  // This algebra is based on the paper
3  // "Cubic curves and cubic surfaces from contact
4  //  points in conformal geometric
5  //  algebra" by Hitzer and Hildenbrand
6
7  // points
8  // e_1 = e1
9  // e_2 = e2
10 // e_{o{k}} = eo{k}
11 // e_{\infty {k}} = ei{k}
12 ei = { 0.5*(ei1 + ei2) }
13 eo = { eo1 + eo2 }
14
15 // need to reference _P(3) as GAALOP's
16 // visualization assumes a 3D space
17 createPoint = {
18     _P(1)*e1 + _P(2)*e2 + 0.5*(_P(1)*_P(1)*ei1
19   + _P(2)*_P(2)*ei2) + _P(1)*_P(2)*ei3
20   + _P(1)*_P(1)*_P(1)*ei4 + _P(1)*_P(1)*_P(2)*ei5
21   + _P(1)*_P(2)*_P(2)*ei6 + _P(2)*_P(2)*_P(2)*ei7
22   + eo() + _P(3)*0
23 }
24
25 // bivectors (E_i, E)
26 E1 = { ei1 ^ eo1 }
27 E2 = { ei2 ^ eo2 }
```

```
28 E3 = { ei3 ^ eo3 }
29 E4 = { ei4 ^ eo4 }
30 E5 = { ei5 ^ eo5 }
31 E6 = { ei6 ^ eo6 }
32 E7 = { ei7 ^ eo7 }
33 E = { ei() ^ eo() }
34
35 // pseudo-scalars
36 Ieps = { e1 * e2 }
37
38 Ii12 = { ei1 * ei2 }
39 Iic = { ei4 * ei5 * ei6 * ei7 }
40 Iib = { ei3 * Iic() }
41 Ii = { Ii12() * Iib() }
42
43 Io12 = { eo1 * eo2 }
44 Ioc = { eo4 * eo5 * eo6 * eo7 }
45 Iob = { eo3 * Ioc() }
46 Io = { Io12() * Iob() }
47
48 Iio = { Ii() ^ Io() }
49
50 eo12T = { eo1 - eo2 }
51 IoT = { eo12T() * Iob() }
52 ei12T = { ei1 - ei2 }
53 IiT = { ei12T() * Iib() }
54
55 I = { Ieps() * Iio() }
56
57 // -----
58 Dual = {
59     -(_P(1) * I()) // a^* = aI^{-1} = a(-I)
60 }
```

At first, the representations of infinity and origin as well as of arbitrary points are defined according to the equations (2) and (24) of [38]. After some definitions of bivectors and pseudoscalars the dual function is defined.

12.2 CUBIC CURVES

Here, we describe how CCGA handles plane cubic curves [38]. A cubic curve in $\mathbb{R}^2$ is formulated as

$$F(x,y) = \mathrm{a}x^3 + \mathrm{b}x^2y + \mathrm{c}xy^2 + \mathrm{d}y^3 + \mathrm{e}x^2 + \mathrm{f}y^2 + \mathrm{g}xy + \mathrm{h}x + \mathrm{i}y + \mathrm{j} = 0. \quad (12.1)$$

The first way to represent a cubic curve in CCGA is constructive by wedging nine contact points together as follows

$$\mathbf{q} = \mathbf{x}_1 \wedge \mathbf{x}_2 \wedge \cdots \wedge \mathbf{x}_9. \quad (12.2)$$

The expression for the dual 1-vector is according to [38]

$$-(2\mathrm{e}\mathbf{e}_{o1} + 2\mathrm{f}\mathbf{e}_{o2} + \mathrm{g}\mathbf{e}_{o3} + \mathrm{a}\mathbf{e}_{o4} + \mathrm{b}\mathbf{e}_{o5} + \mathrm{c}\mathbf{e}_{o6} + \mathrm{d}\mathbf{e}_{o7}) + \mathrm{h}\mathbf{e}_1 + \mathrm{i}\mathbf{e}_2 - \mathrm{j}\mathbf{e}_\infty. \quad (12.3)$$

This is computed in Listing 12.3 for a simple curve with a = 1, e = 1, f = −1.

Listing 12.3: CCGA example for the computation of a simple cubic curve.

```
1  // CCGA cubic curve
2
3  // simple cubic (a = 1, e = 1, f = -1)
4  V = -(2*eo1 -2*eo2 + eo4);
5
6  :Black;
7  :V;
```

Its visualization can be seen in Fig. 12.1. Please notice that plane curves are visualized by Vis3D with an extension into the third space dimension as some kind of cylinder.

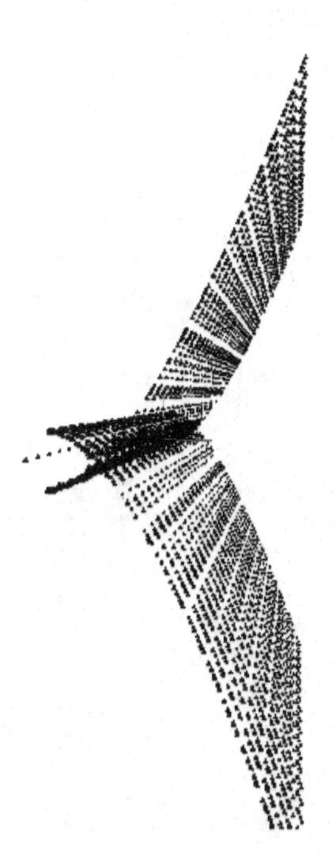

Figure 12.1 Vis3D visualization of a simple cubic curve.

CHAPTER 13

GAALOPWeb for GAPP

CONTENTS

Geometric Algebra has an inherent potential for parallelization. This can be very well seen in an intermediate language of GAALOP called GAPP. This chapter describes GAALOPWeb for GAPP which is very well suitable for hardware implementations. This can be seen in Chapt. 14 showing how *GAALOPWeb for GAPPCO* is dealing with the Geometric Algebra hardware design GAPPCO [35]. Chapt. 15 describes a new hardware design called GAPPCO II.

13.1 THE REFLECTOR EXAMPLE

As an example, we use the reflection of a sphere in a plane of [35]. The reflection operation is the most basic transformation in Conformal Geometric Algebra, since all the conformal transformations can be composed based on it. Rotations and translations, for instance, consist of two consecutive reflections related to two specific planes.

Figure 13.1 shows the reflection of a (green) sphere related to the (red) plane, which is resulting in the reflected (yellow) sphere. The corresponding algorithm is described in Listing 13.1.

Listing 13.1: GaalopScript for the computation of the reflection of an object a at a plane m

```
1 // Code to optimize
2 a = a1*e1+a2*e2+a3*e3+a4*einf+ e0;
3 m = m1*e1+m2*e2+m3*e3+m4*einf;
4 ?Dotproduct = a.m;
5 a_par = 2*(Dotproduct)*m;
6 ?a_Refl = 0.5*(a - a_par);
```

This GAALOPScript first defines the objects a and m. The object a is either a point or a sphere to be reflected at m which is a plane with normal vector (m1,m2,m3)

DOI: 10.1201/9781003139003-13

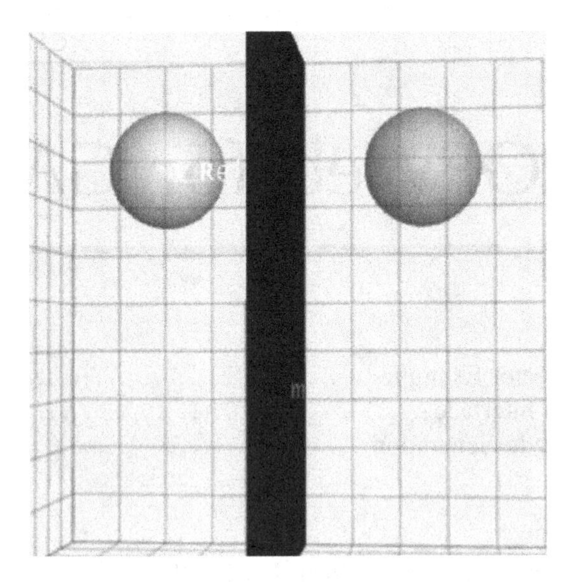

Figure 13.1 Reflection of a sphere according to the algorithm of Listing 13.1.

and m4 as the distance to the origin (please refer to Sect. 2.2). The reflection operation is defined according to its description in [22].

The question marks indicate the multivectors to be computed explicitly (either as an intermediate or as a final result). Note: the multiplication of the result by 0.5 is done, because the resulting computations become easier (and in Conformal Geometric Algebra the multiplication with a scalar does not change the geometric object).

13.2 THE WEB INTERFACE

Using additionally the GAALOPScript code of the following two listings, the visualization of Figure 13.1 can be generated.

Listing 13.2: Variable assignments for the visualization of the reflection of an object a (point or sphere) related to a plane m

```
1 // Variable assignments
2 a1 = 0.5;
3 a2 = 0.5;
4 a3 = 0.5;
5 a4 = 0.4;
6 m1 = 1;
7 m2 = 0;
8 m3 = 0;
9 m4 = 0;
```

Listing 13.3: Multivectors to be visualized for the visualization of the reflection of an object *a* (point or sphere) related to a plane *m*

```
1 // Multivectors to be visualized
2 :Green;
3 :a;
4 :Red;
5 :m;
6 :Yellow;
7 :a_Refl;
```

Here the original sphere is visualized in green, the plane in red and the reflected sphere in yellow. Figure 13.2 shows a screenshot of GAALOPWeb for GAPP integrating the three listings 13.1, 13.2 and 13.3.

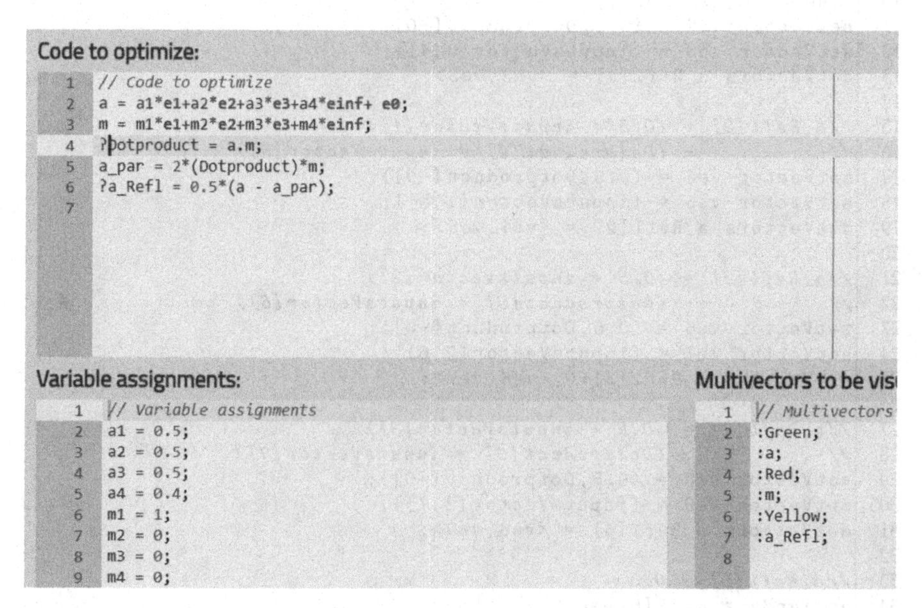

Figure 13.2 Screenshot of "GAALOPWeb for GAPP".

13.3 GAPP CODE GENERATION

We use GAALOPWeb for the symbolic optimization of Geometric Algebra algorithms, resulting finally in the optimized GAPP code. We describe this process for GAPPCO based on the example of Sect. 13.1.

The GAPP language as an IL (intermediate language) of GAALOPWeb describes the general structure of the computations after the general optimization process. As described in the book "Foundations of Geometric Algebra Computing", Geometric Algebra algorithms with all kind of products of multivectors always have the same principle structure. This is described in the GAPP instruction set of Table 14.1 of [29]. If, for instance, divisions or square roots are needed, the instruction set has to be extended (see [62]).

The reflection algorithm of Listing 13.1 does not use any extended operations such as divisions and square roots. This is why the corresponding GAPP code uses the standard GAPP instruction set as presented in Listing 13.4.

Listing 13.4: Resulting GAPP code of Listing 13.1.

```
1  assignInputsVector inputsVector = [a1,a2,a3,a4,m1,m2,m3,m4];
2
3  resetMv Dotproduct[32];
4  setVector ve0 = {inputsVector[-7,2,1,0]};
5  setVector ve1 = {1.0,inputsVector[6,5,4]};
6  dotVectors Dotproduct[0] = <ve0,ve1>;
7
8  //a_Refl[1] = (0.5 * inputsVector[0])
9  //          - (Dotproduct[0] * inputsVector[4])
10 resetMv a_Refl[32];
11 setVector ve2 = {0.5,Dotproduct[-0]};
12 setVector ve3 = {inputsVector[0,4]};
13 dotVectors a_Refl[1] = <ve2,ve3>;
14
15 //a_Refl[2] = (0.5 * inputsVector[1])
16 //          - (Dotproduct[0] * inputsVector[5])
17 setVector ve4 = {0.5,Dotproduct[-0]};
18 setVector ve5 = {inputsVector[1,5]};
19 dotVectors a_Refl[2] = <ve4,ve5>;
20
21 //a_Refl[3] = (0.5 * inputsVector[2])
22 //          - (Dotproduct[0] * inputsVector[6])
23 setVector ve6 = {0.5,Dotproduct[-0]};
24 setVector ve7 = {inputsVector[2,6]};
25 dotVectors a_Refl[3] = <ve6,ve7>;
26
27 //a_Refl[4] = (0.5 * inputsVector[3])
28 //          - (Dotproduct[0] * inputsVector[7])
29 setVector ve8 = {0.5,Dotproduct[-0]};
30 setVector ve9 = {inputsVector[3,7]};
31 dotVectors a_Refl[4] = <ve8,ve9>;
32
33 //a_Refl[5] = 0.5
34 assignMv a_Refl[5] = [0.5];
```

First of all, the 8 input values of the program (a1,a2,a3,a4,m1,m2,m3 and m4) are assigned to the *inputsVector* (vector with all the scalar input values). Then, the coefficients of the multivectors *Dotproduct* and `a_Refl` are computed. From the multivector *Dotproduct* only the index 0 is needed. It is the dot product of the two vectors ve0 and ve1 consisting of entries from the inputsVector as well as the value 1.0 as a constant. Then the entries 1, 2, 3, 4 and 5 of the multivector `a_Refl` are computed. The first entry, for instance, is the dot product of the vectors ve2 and ve3 consisting of input values, constants and the result of the negated Dotproduct[0] (because of consistency reasons, −Dotproduct[0] is written Dotproduct[−0] in the GAPP language).

CHAPTER 14

GAALOPWeb for GAPPCO

CONTENTS

During the last few years there has been an increasing interest in using Geometric Algebra in engineering applications, especially in computer graphics, computer vision and robotics. Geometric Algebra is a very powerful mathematical language combining both geometric intuitivity and the potential of high runtime performance of the implementations. This makes it very promising to take Geometric Algebra as a Domain-Specific Language (DSL) and map it to very efficient implementations on FPGAs.

A good way of losing the high complexity of Geometric Algebra before going to the real computing device is to precompile Geometric Algebra algorithms based on GAALOP (see [29] and [30]). What remains after its optimization process are mainly parallel computations of multivector coefficients each consisting of sums of products, which are again efficiently to be parallelized.

GAALOP already supports optimized hardware (HW) generation [64] based on a backend for Verlilog. But, with every new algorithm the FPGA has to be completely reprogrammed. The solution of the paper "GAPPCO: an Easy to Configure Geometric Algebra Co-processor Based on GAPP Programs" [35] is more far-reaching. Its design combines the GAALOP technology with new developments in fixed HW solutions for Geometric Algebra, especially the *ConformalALU*of [22], implementing conformal transformations in hardware. We use this solution as an example for GAPPCO, a general co-processor design based on GAPP layer of GAALOPWeb.

DOI: 10.1201/9781003139003-14

In order to introduce GAALOP to the hardware community, our requirement was to extend GAALOP in a way to make it as easy as possible to handle for FPGA developers/users. This is why we developed a new web-based tool for GAPPCO solutions.

The advantages of the web-based solution compared to [35] are that

- it can be used from everywhere without any need of installing a software.

- one continuous compilation process (no need for different tools for different stages)

- support of the geometric intuitivity of Geometric Algebra by an integrated visualization

For our example, we use Conformal Geometric Algebra according to Sect. 2.2, since it is very well suitable to realize engineering applications. This is primarily because of its easy handling of geometric objects such as spheres, planes and lines. For more details about Conformal Geometric Algebra see [29].

14.1 GAPPCO IN GENERAL

GAPPCO is a co-processor design combining both the advantages of an optimizing software with a configurable hardware able to implement arbitrary Geometric Algebra algorithms. The idea is to have a fixed hardware, easily and fast to be configured for different algorithms. Compared to standard hardware architectures it makes use of variable-size vectors and fast register access. While the GAPPCO design is a design for reconfigurable hardware, it is flexible in the sense that it can be used for specific hardware devices such as ASICs or SOCs in the future.

GAPPCO consists of one or more GAPP units. Each GAPP unit is able to realize small GAPP programs (as, for instance, the GAPP program of Listing 13.4) in parallel. The host interface is responsible for the data communication between host and

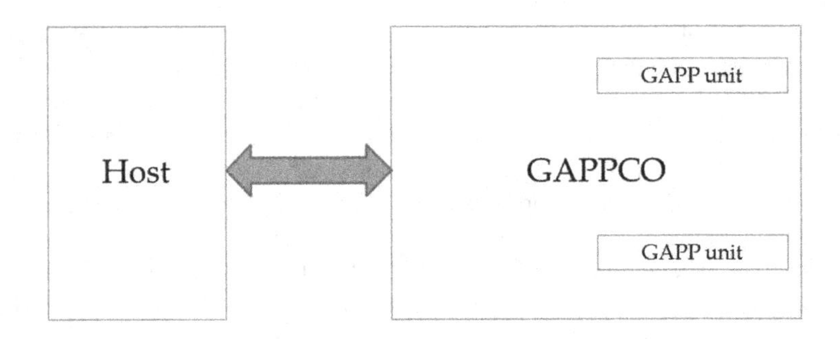

Figure 14.1 GAPPCO co-processor with host interface.

GAPPCO and for the configuration of GAPPCO (see Fig. 14.1).

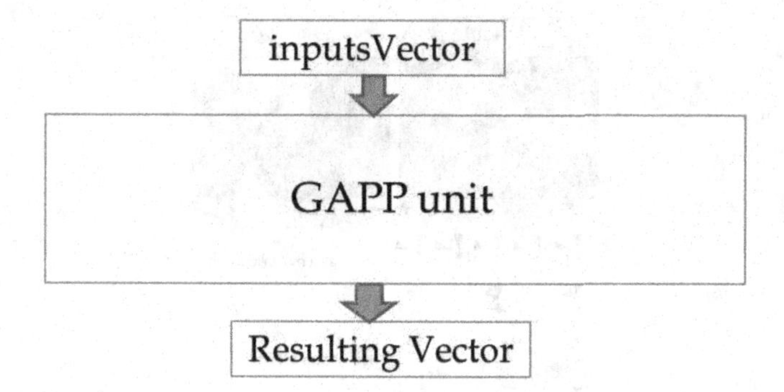

Figure 14.2 GAPP unit processing an input Vector (vector with all the scalar input values) to resulting vectors.

As soon as a GAPP unit is configured, inputsVectors can be received from the host and resulting vectors can be sent to the host (see Fig. 14.2). For runtime performance purposes, the architecture of GAPP units is pipelined as much as possible.

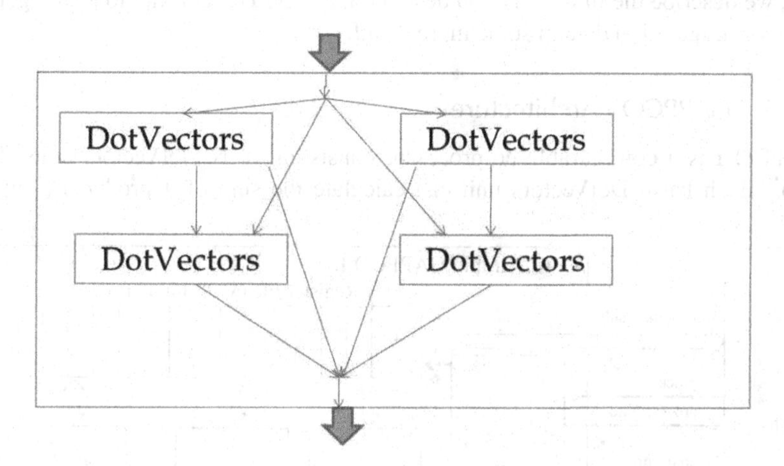

Figure 14.3 GAPP unit consisting of 2 levels of DotVectors units.

A GAPP unit consists of a network of one or more DotVectors units organized in one or more levels (see Fig. 14.3). The connections have to be configured before runtime based on the specific configuration data (see Sect. 14.2.3). Each DotVectors unit is responsible for the computation of one coefficient of one multivector. There is an implicit parallelism in this computation according to Fig. 14.4. The multiplications of each of the vector elements can be done in parallel as well as parts of the additions (see Fig. 14.4). The GAPP listing 13.4 is restricted to the number of two vectors to be multiplied, while GAPP, generally, is supporting a higher number of vectors to be multiplied.

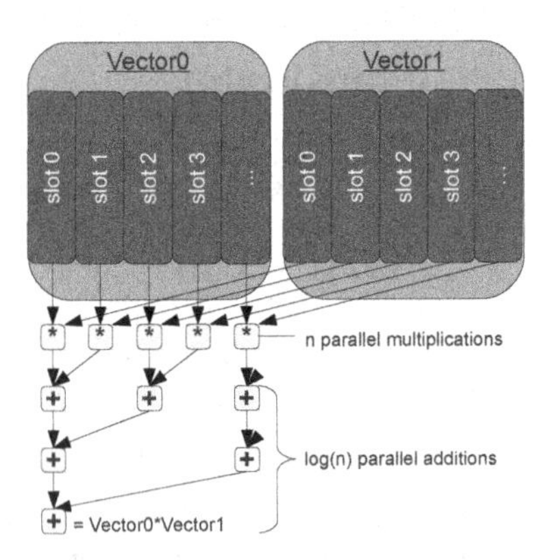

Figure 14.4 Parallel dot product of two n-dimensional vectors Vector0 and Vector1 (n parallel products followed by log(n) parallel addition steps).

14.2 GAPPCO I

Here, we describe the first GAPPCO design (GAPPCO I) according to [35] together with its configuration data in some more detail.

14.2.1 GAPPCO I Architecture

GAPPCO I is a configurable co-processor consisting of N DotVectors units (Fig. 14.5). Each basic DotVectors unit can calculate the sum of 4 products (4-width

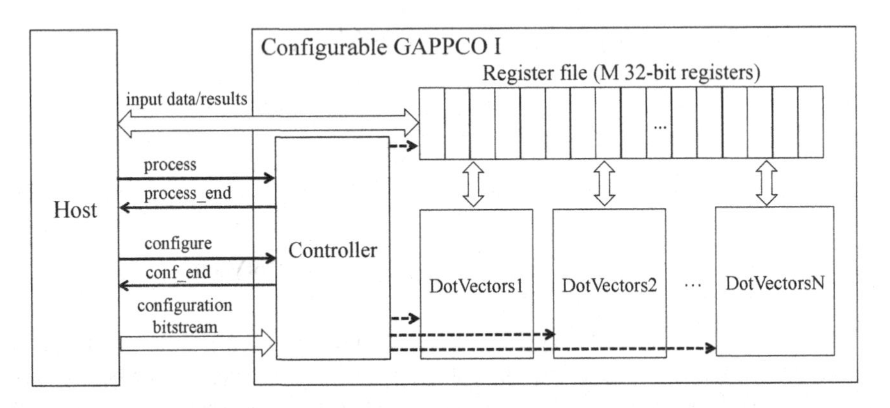

Figure 14.5 Configurable GAPPCO I block diagram.

DotVectors unit). The block diagram of each DotVectors unit is depicted in Fig. 14.6. It is composed of 4 multiplier units (MULT11, MULT12, MULT13 and MULT14) and 3 adder units (ADD11, ADD12 and ADD13). Each DotVectors unit can be configured as one 4-width DotVectors unit or two 2-width DotVectors units. To make the unit configurable, considering the first level of adders, the output of each adder unit can be either provided as result or used as input of a further adder unit. A demultiplexer unit is used for this purpose. The enable inputs of each demultiplexer are part of the configuration data provided by the GAPPCO configurator program on the host side.

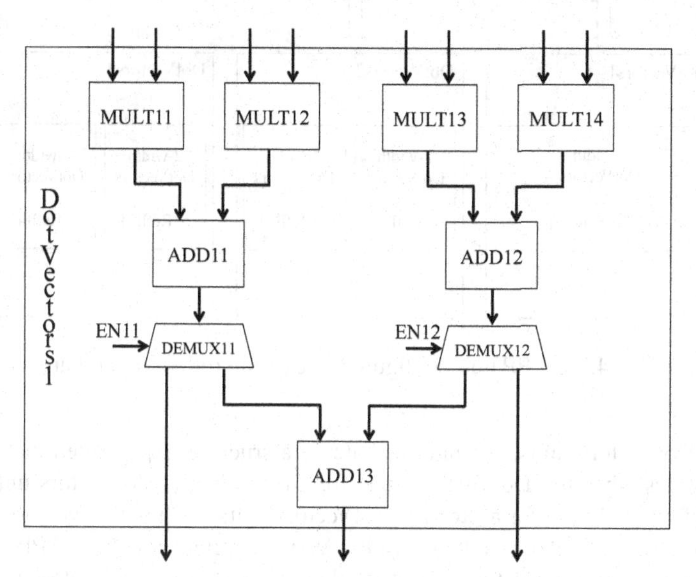

Figure 14.6 DotVectors1 block diagram.

This first GAPPCO design uses basic 4-width DotVectors units since this width is sufficient to support reflection operations and therefore the fundamental conformal geometric operations. As described in [22], each conformal geometric operation can be obtained as multiple consecutive reflection operations.

The number N of DotVectors units depends on the resource availability on the target FPGA device. As shown in Fig. 14.5, GAPPCO I also includes a controller unit and a register file consisting of M 32-bit registers to store input data (inputsVector and constants) and output results (both intermediate and final) of the DotVectors units.

14.2.2 The Compilation Process

In a nutshell, we use GAALOPWeb for the symbolic optimization of Geometric Algebra algorithms, resulting finally in the specific configuration data of GAPPCO. We describe this process for GAPPCO based on the example of Sect. 13.1. At first we compile to the *GAPP* (Geometric Algebra Parallelism Programs) layer according to Sect. 13.3.

For the generation of the GAPPCO configuration data, the particular GAPP layer data has to be translated in order to configure GAPPCO. Fig. 14.7 shows the GAPP unit configured to execute reflection operations. As required by the reflection operation, three basic DotVectors units are used: the first one is configured as one 4-width DotVectors unit, while the second one and the third one are both configured as two

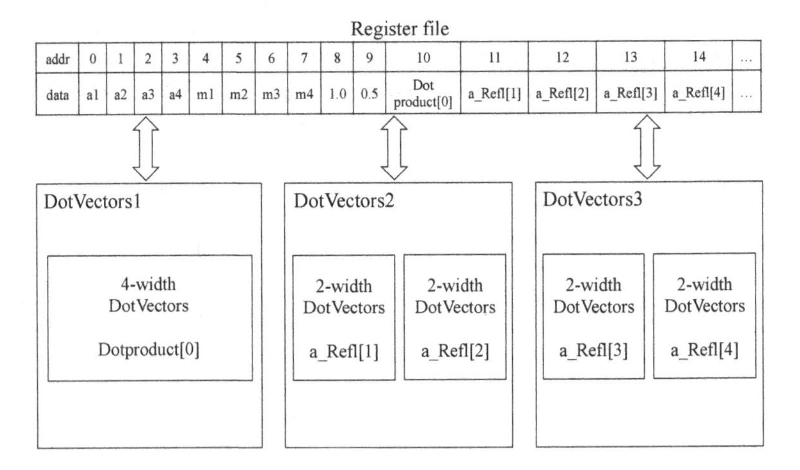

Figure 14.7 GAPP unit configured to execute reflection operations.

2-width DotVectors units. An intermediate data sructure is presented in Fig. 14.8. It mainly describes the DotVectors units for *Dotproduct* (1 dotVectors unit with 2 vectors of width 4) and for `a_Refl` (4 dotVectors units, each with 2 vectors of width 2) with their intermediate or final results. Very important for the GAPP unit configuration is the routing list assigning numbers to input values, constant values and intermediate values. These numbers are used for the input definitions of the DotVectors units. In our example, the needed constants are 0.5 and 1.0. The routing list is the base for the register file as one important component of the GAPPCO I design presented in Sect. 14.2.

Finally, the concrete bitstream for the configuration of GAPPCO has to be computed. You can see the result for our specific reflector example in Table 14.3.

14.2.3 Configuration Phase

Before runtime, the configurator program on the host side provides the configuration bitstream needed to configure DotVectors units. When the controller receives the "configure" command, it configures GAPPCO I according to the configuration bits and sends the "conf˙end" status signal to the host (see Fig. 14.5).

As shown in Fig. 14.6, each basic DotVectors unit consists of 4 multiplier units, 3 adder units, and 2 demultiplexer units. The first index of each multiplier, adder or demultiplexer unit indicates the number of the DotVectors unit, while the second index specifies the number of the multiplier, adder or demux unit. For each 4-width DotVectors unit, the configuration bitstream is composed of:

enable inputs of demultiplexer units

addresses of input operands of each multiplier unit

signs of input operands of each multiplier unit

```
inputsVector:  a1  a2  a3  a4  m1  m2  m3  m4

Routing list:
0: inputsVector[0]
1: inputsVector[1]
2: inputsVector[2]
3: inputsVector[3]
4: inputsVector[4]
5: inputsVector[5]
6: inputsVector[6]
7: inputsVector[7]
8: 1.0
9: 0.5
10: Dotproduct[0]


DotVectors unit
input:  -7  2  1  0
input:  8  6  5  4
Level 1 result: Dotproduct[0]

DotVectors unit
input:  9  -10
input:  0  4
Level 0 result: a_Refl[1]

DotVectors unit
input:  9  -10
input:  1  5
Level 0 result: a_Refl[2]

DotVectors unit
input:  9  -10
input:  2  6
Level 0 result: a_Refl[3]

DotVectors unit
input:  9  -10
input:  3  7
Level 0 result: a_Refl[4]
```

Figure 14.8 Internal configuration data structure for GAPPCO based on the reflection example.

address (addresses) of output result (results)

result/results type (intermediate or final)

As an example, the configuration bits for the DotVectors1 unit in Fig. 14.6 are reported in Table 14.1. The enable bit values of the demultiplexer units (EN11 and EN12) specify if the DotVectors unit has to be configured as one 4-width DotVectors unit (EN11=EN12=0) or two 2-width DotVectors units (EN11=EN12=1). Furthermore, for each multiplier unit, the configuration bitstream specifies the signs of the input operands (e.g. MULT11_sign1 and MULT11_sign2) as well as the addresses of the registers (within the register file) where the input operands have to be read (e.g. MULT11_addr1 and MULT11_addr2). If the operand sign is negative (sign bit = 1), a sign changing operation will be performed before the operand is sent to the multiplier unit to be processed. Regarding the register addresses, a register file composed of 32 32-bit registers has been considered for this first design and therefore each address field in the configuration bitstream is composed of 5 bits. Finally, for each configured DotVectors unit, the configuration data specifies the address of the register where the output result has to be written back (e.g. RESULT11_addr) as well as the result type (intermediate calculation or final result). If the DotVectors unit is configured as two 2-width DotVectors units, two results will be provided (RESULT11 and RESULT12). The configuration data are read by the controller unit that configures accordingly the configurable DotVectors units.

TABLE 14.1 Configuration bits for DotVectors1 of Fig. 14.6. (A register file composed of 32 32-bit registers has been considered for this first design and therefore 5-bit addresses are needed.)

Field	**N. of bits**
EN11	1
EN12	1
MULT11_addr1	5
MULT11_sign1	1
MULT11_addr2	5
MULT11_sign2	1
MULT12_addr1	5
MULT12_sign1	1
MULT12_addr2	5
MULT12_sign2	1
MULT13_addr1	5
MULT13_sign1	1
MULT13_addr2	5
MULT13_sign2	1
MULT14_addr1	5
MULT14_sign1	1
MULT14_addr2	5
MULT14_sign2	1
RESULT11_addr	5
RESULT11_type (intermediate or final)	1
RESULT12_addr (optional)	5
RESULT12_type (optional)	1

This first GAPPCO design is restricted to two types of results, namely, intermediate and final, as needed in the examples chosen for this first proof-of-concept. Therefore, one configuration bit is used to specify the result type (0 for final results and 1 for intermediate results). However, more complex GA applications need different levels of intermediate results. Final computations need intermediate results of level 1, level 1 computations have to wait for intermediate results of level 2, level 2 computations have to wait for intermediate results of level 3, and so on.

GAPP units of different size each containing a different number of variable-width DotVectors units can be configured within the GAPPCO co-processor. The entire configuration bitstream of GAPPCO I will be composed of the fields reported in Table 14.2. The first 4-bit field of the bitstream specifies the number of GAPP units to be configured within the co-processor (e.g. r). Furthermore, for each GAPP unit, the configuration bitstream specifies the number of basic 4-width DotVectors units that are needed to obtain the required GAPP unit as well as the configuration bits needed to configure each basic DotVectors unit. The configuration data for each basic DotVectors unit are structured as reported in Table 14.1. Using the configuration bits listed in Table 14.2, we can configure a GAPPCO I co-processor composed of **up to 16 GAPP** units each consisting of **up to 16 4-width DotVectors units**.

TABLE 14.2 Configuration Bitstream of GAPPCO I

Field	N. of bits
N. of GAPP units	4
N. of 4-width DotVectors units for GAPP unit 1	4
Configuration bits for DotVectors1 of GAPP unit 1	as listed in Table 14.1
Configuration bits for DotVectors2 of GAPP unit 1	"
...	...
Configuration bits for DotVectors*n* of GAPP unit 1	"
N. of 4-width DotVectors units for GAPP unit 2	4
Configuration bits for DotVectors1 of GAPP unit 2	as listed in Table 14.1
Configuration bits for DotVectors2 of GAPP unit 2	"
...	...
Configuration bits for DotVectors*m* of GAPP unit 2	"
...	...
N. of 4-width DotVectors units for GAPP unit *r*	4
Configuration bits for DotVectors1 of GAPP unit *r*	as listed in Table 14.1
Configuration bits for DotVectors2 of GAPP unit *r*	"
...	...
Configuration bits for DotVectors*p* of GAPP unit *r*	"

For our reflector example the configuration bitstream is reported in Table 14.3. It is based on the configuration data format presented in Fig. 14.8.

14.2.4 Runtime Phase

As shown in Fig. 14.5, at runtime, when the host sets the "process" control signal, GAPPCO I starts input data processing, and, after completion, sets the "process_end" status signal.

GAPPCO I is based on a pipelined architecture.

System operation phases at runtime are as follows:

1. The host writes input data (inputsVector and constantsVector) to the register file
2. The co-processor reads input data from the register file
3. The co-processor executes operations
4. The co-processor writes results (both intermediate and final) back to the register file
5. The host reads results from the register file

The controller unit supervises the DotVectors units operation. Each DotVectors unit reads its input data from the proper registers of the register file (as set during the configuration phase) and writes the result (results) to the proper register (registers) of the register file. If this result is an intermediate result, it will be further processed by other DotVectors units.

14.3 THE WEB INTERFACE

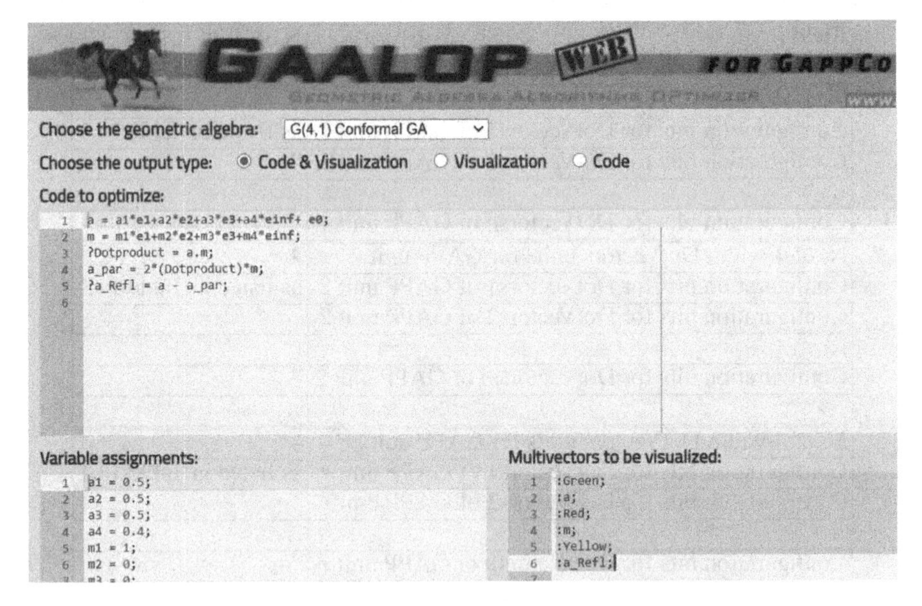

Figure 14.9 Screenshot of "GAALOPWeb for GAPPCO".

Figure 14.9 shows the reflector example integrated in "GAALOPWeb for GAPPCO". The user is able to select a specific Geometric Algebra and whether he is interested in code and/or visualization. It offers separate text areas for

1. Variable assignments
2. Code to optimize
3. Multivectors to be visualized

If you are only interested in the code (configuration data for GAPPCO), only the text area 2. is needed. If you are also interested in a visualization of your algorithm, both the text areas 1. and 3. are needed.

Please refer to their contents for our reflector example as follows:

1. Variable assignments: Listing 13.2
2. Code to optimize: Listing 13.1
3. Multivectors to be visualized: Listing 13.3

TABLE 14.3 Configuration Bitstream for GAPP unit Realizing reflector Functionality

Fields	**Values**
N. of GAPP units	1
N. of 4-width DotVectors units for GAPP unit 1	3
Configuration bits for DotVectors1	
EN11/EN12	0/0
MULT11_addr1/MULT11_sign1 (0 for +, 1 for −)	7/1
MULT11_addr2/MULT11_sign2	8/0
MULT12_addr1/MULT12_sign1	2/0
MULT12_addr2/MULT12_sign2	6/0
MULT13_addr1/MULT13_sign1	1/0
MULT13_addr2/MULT13_sign2	5/0
MULT14_addr1/MULT14_sign1	0/0
MULT14_addr2/MULT14_sign2	4/0
RESULT11_addr/RESULT11_type (0 for final, 1 for intermediate)	10/1
RESULT12_addr (optional)/RESULT12_type (optional)	-/-
Configuration bits for DotVectors2	
EN21/EN22	1/1
MULT21_addr1/MULT21_sign1 (0 for +, 1 for −)	9/0
MULT21_addr2/MULT21_sign2	0/0
MULT22_addr1/MULT22_sign1	10/1
MULT22_addr2/MULT22_sign2	4/0
MULT23_addr1/MULT23_sign1	9/0
MULT23_addr2/MULT23_sign2	1/0
MULT24_addr1/MULT24_sign1	10/1
MULT24_addr2/MULT24_sign2	5/0
RESULT21_addr/RESULT21_type (0 for final, 1 for intermediate)	11/0
RESULT22_addr (optional)/RESULT22_type (optional)	12/0
Configuration bits for DotVectors3	
EN31/EN32	1/1
MULT31_addr1/MULT31_sign1 (0 for +, 1 for −)	9/0
MULT31_addr2/MULT31_sign2	2/0
MULT32_addr1/MULT32_sign1	10/1
MULT32_addr2/MULT32_sign2	6/0
MULT33_addr1/MULT33_sign1	9/0
MULT33_addr2/MULT33_sign2	3/0
MULT34_addr1/MULT34_sign1	10/1
MULT34_addr2/MULT34_sign2	7/0
RESULT31_addr/RESULT31_type (0 for final, 1 for intermediate)	13/0
RESULT32_addr (optional)/RESULT32_type (optional)	14/0

CHAPTER 15

GAPPCO II

CONTENTS

GAPPCO I according to Sect. 14.2 is an appropriate concept for small Geometric Algebra algorithms. It is restricted to two levels (algorithm steps) and DotVectors units with 2 vectors of max. size 4.

15.1 THE PRINCIPLE

Figure 15.1 shows the difference between the GAPPCO I and GAPPCO II designs. While the GAPPCO I design consists of N parallel DotVectors units, the GAPPCO

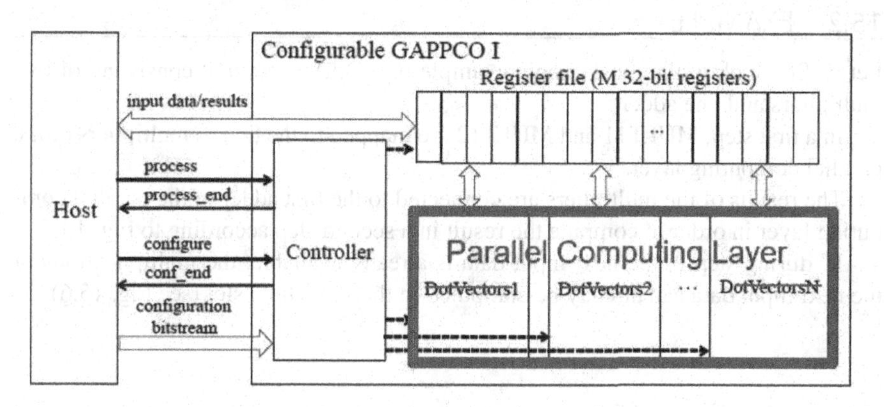

Figure 15.1 The general GAPPCO II design.

II design consists of one small parallel computing layer according to Fig. 15.2. Many multipliers, adders, ... are working in parallel. The controller is responsible for the correct distribution of inputs and outputs to the registers of the register file. In principle, its task is to map a DotVectors unit to a sequence of parallel computations based on the parallel computing layer (see Fig. 15.3).

DOI: 10.1201/9781003139003-15

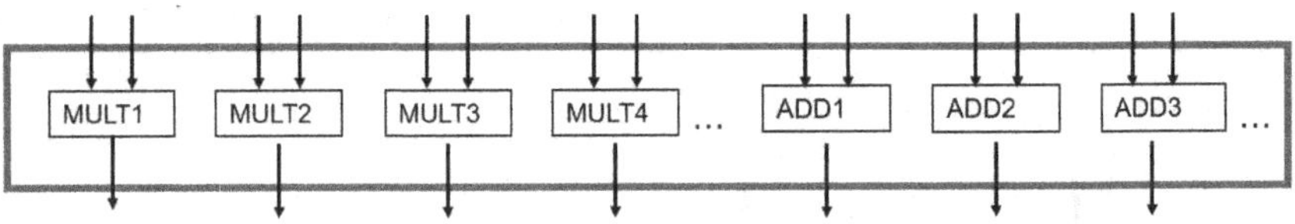

Figure 15.2 The parallel computing layer.

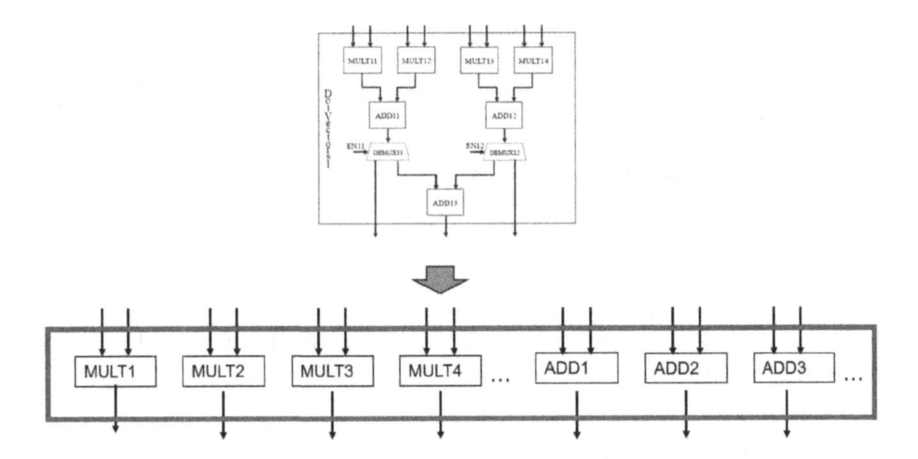

Figure 15.3 Mapping of a DotVectors unit into the GAPPCO II design.

15.2 EXAMPLE

Let us first look to the most simple example of a DotVectors unit consisting of two multipliers and one adder.

In a first step, MULT11 and MULT12 are mapped to the first to multipliers of the parallel computing layer.

The results of the multipliers are connected to the first adder of the parallel computing layer in order to compute the result in a second step according to Fig. 15.5.

If, during step 2, the next input data is already available, the multiplications of the next input data can already be computed in the same time slot (see Fig. 15.6).

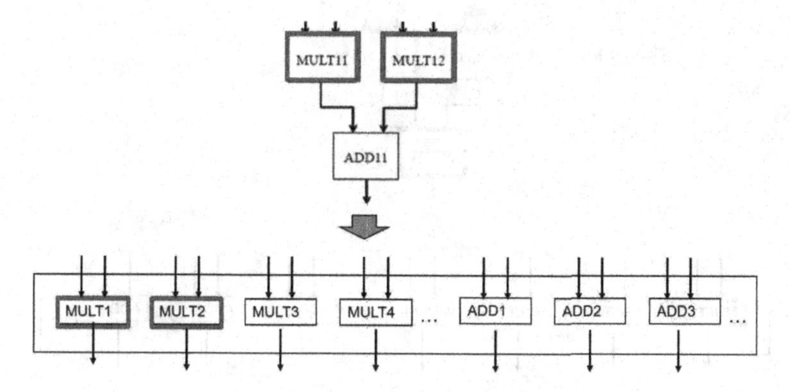

Figure 15.4 Step 1.

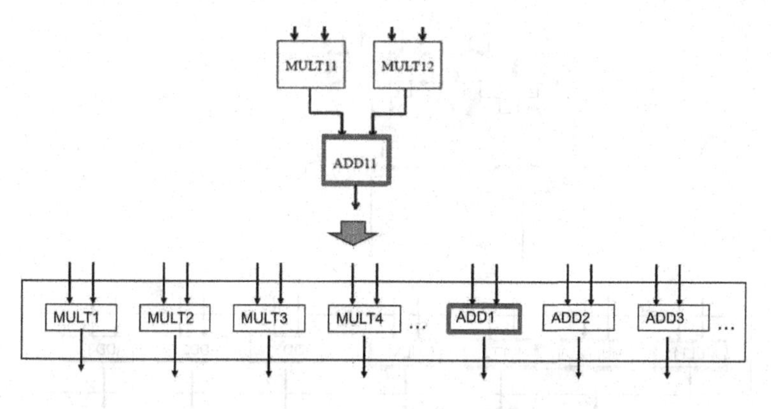

Figure 15.5 Step 2.

Now let us look into a DotVectors example with 4 multipliers (the maximum in the GAPPCO I design).

In a first step, the first four multipliers are used according to Fig. 15.7.

In a second step the first two adders are used and at the same time the four multiplications of the next input data are computed according to Fig. 15.8.

And, in the third step the result of the first input data is computed parallel to the additions of the second input data and the multiplications of the third input data (see Fig. 15.8).

15.3 IMPLEMENTATION ISSUES

For the implementation of GAPPCO II the following issues are important

- the size of the register file

- the number of computing units

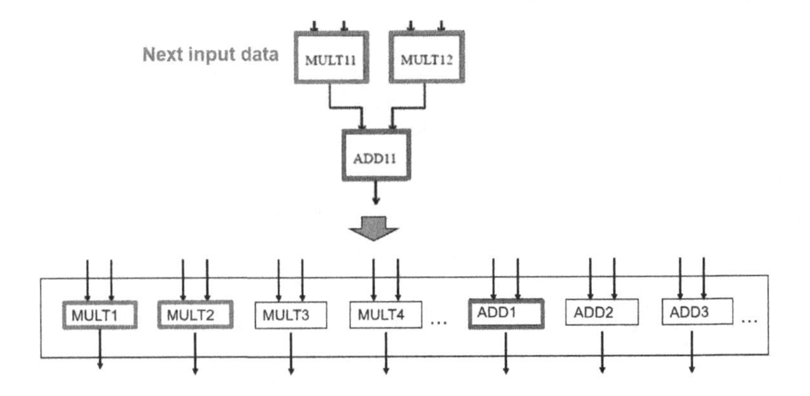

Figure 15.6 Step 2A.

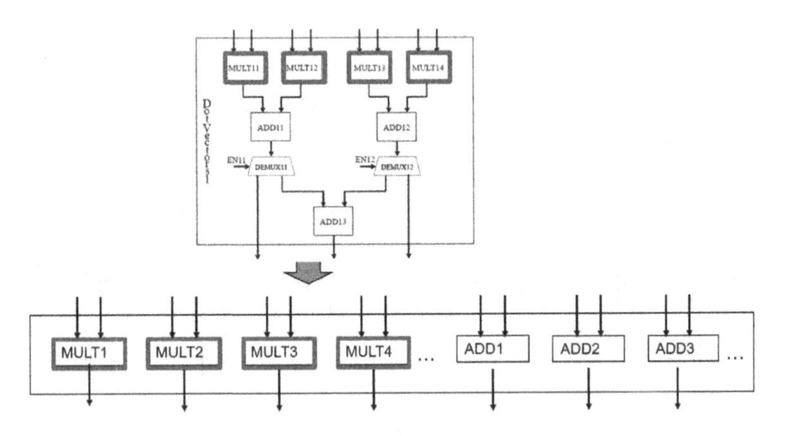

Figure 15.7 Step 1.

- the handling of their different processing times

Appropriate values could be for instance

- 128 registers

- 96 adders

- 128 multipliers

- 1 divider

- 1 square root unit

The operations of the parallel computing layer have different processing times as indicated in Fig. 15.10. Adders are the fastest, multipliers the next (approx. three times slower), then divisions (approx. six times slower) and square roots are taking

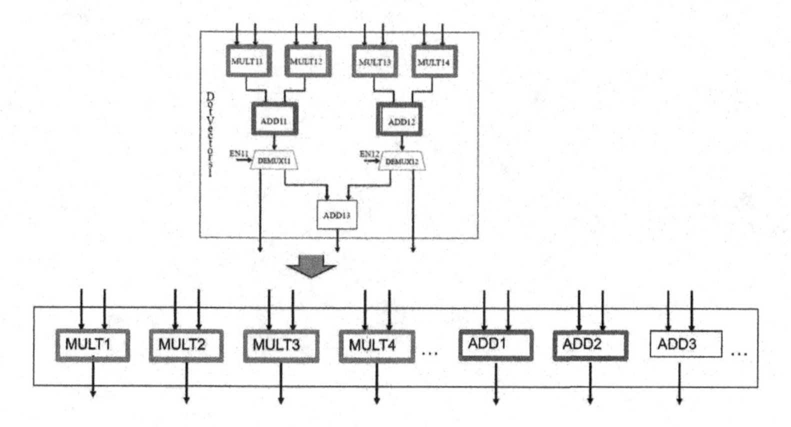

Figure 15.8 Step 2.

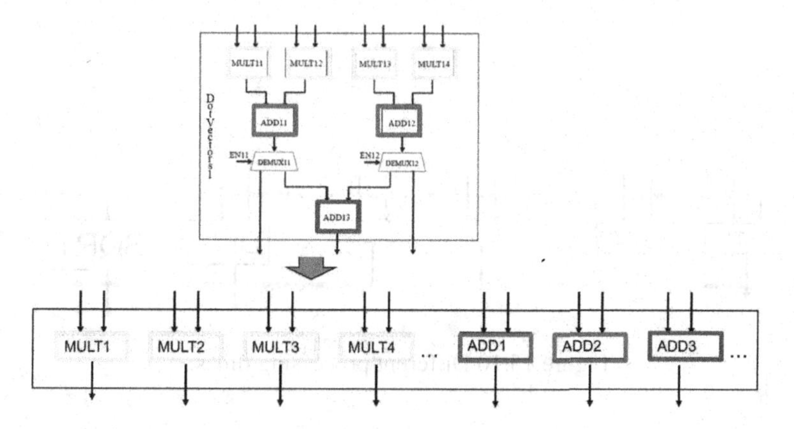

Figure 15.9 Step 3.

the most time (approx. 20 times slower). The controller according to Fig. 15.1 is responsible for the correct activation of the computing units according to their number and processing times.

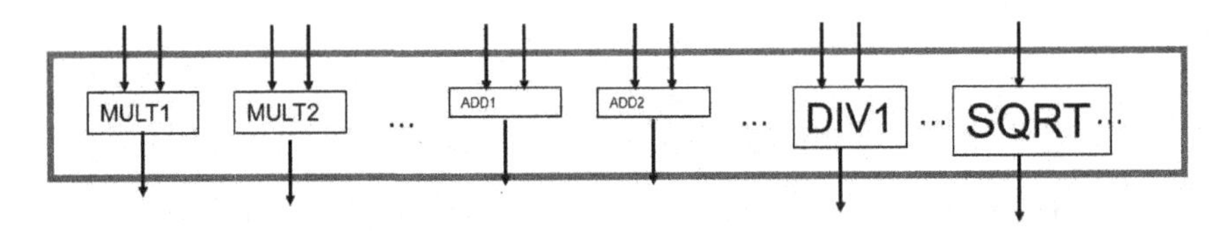

Figure 15.10 Different processing times.

CHAPTER 16

Introduction to Quantum Computing

CONTENTS

In this chapter, we introduce quantum computing according to [36] in preparation of the next chapter where we introduce a specific Geometric Algebra to handle it. We will see that the operations of quantum computing can be understood as algebraic transformations such as reflections or rotations of a so-called state vector. As example computing gate we use the NOT-operation, the Hadamard transform as well as the CNOT-operation.

16.1 COMPARING CLASSIC COMPUTERS WITH QUANTUM COMPUTERS

The computers that we use today are deterministic devices with memory and register units composed of unambiguous bits. Each bit can either be entirely in a state of 0 or, oppositely, in a state of 1. They work in the classic domain as each switching operation that changes state 0 to state 1 (or the other way round), still requires an enormous number of charge carriers interacting even with the miniaturization achieved today.

Thus, stepping up the performance of these classic computers necessitates a reduction in the size of their components. It may, therefore, be expected that soon the threshold to quantum mechanics will be achieved once the components become so small that, rather than by the classic laws, they will behave erratically by the laws of quantum mechanics.

DOI: 10.1201/9781003139003-16

One of the possible solutions to this dilemma consists of implementing the possibilities of quantum computing. The prospects concerning further miniaturization are bright: So the performance of a quantum computer using a register with 64 qubits, see page 8 of [44], would correspond to the performance of a classic computer with surface covering several thousands of earth globes. Calculating on such a quantum computer thus represents an extremely efficient way of computing. In quantum computing, this is achieved by specifically coding the information. Whereas each bit of a classic computer can assume both states $|0\rangle$and $|1\rangle$with the likelihoods of either 100% or 0%, the possible states $|0\rangle$or $|1\rangle$of a quantum computer are governed by probabilities lying anywhere between 0% and 100%.

16.2 DESCRIPTION OF QUANTUM BITS

A quantum computer can simultaneously compute and work with the superposition of all possible classic states. Thus, a myriad of states are modified in a single computing step.

In explaining this fact, the question of what is a state, therefore, plays a key role. As a matter of fact, a quantum mechanics state describes a special property of the particle in question such as the spin of an electron. Now, a spin can be directed either downwards (spin down), which corresponds to the quantum bit state of $|0\rangle$or upwards (spin up) corresponding to the quantum bit state of $|1\rangle$. Or else, the electron can be in a superposition of

$$|\Psi\rangle = a_0|0\rangle + a_1|1\rangle, \tag{16.1}$$

where $a_0, a_1 \in \mathbb{C}$. This is the usual representation of a superposition of the two states $|0\rangle$and $|1\rangle$in the conventional notation (see, e.g., [40], [55]) without using Geometric Algebra. Here the pre-factors or coefficients a_0 and a_1 represent the amplitudes of the wave function and are subject to the normalization condition as complex numbers

$$|a_0|^2 + |a_1|^2 = a_0 a_0^* + a_1 a_1^* = 1. \tag{16.2}$$

$|a_0|^2$ is the probability that, in a measurement, the quantum bit indicates the state $|0\rangle$(e.g., spin down), while $|a_1|^2$ is the probability that the quantum bit will be found in the state $|1\rangle$(e.g., spin up).

If the states $|0\rangle$and $|1\rangle$are identified with the space-like basis vectors[1] σ_0 and σ_1 that satisfy the normalization condition

$$\sigma_0^2 = \sigma_1^2 = 1, \tag{16.3}$$

it is possible to define the wave function (16.1) in the Complex Geometric Algebra $\mathscr{C}l(2,\mathbb{C})$ as a location vector r in this space:

$$\psi = a_0\sigma_0 + a_1\sigma_1. \tag{16.4}$$

This geometric-algebraic representation of the state of a quantum bit was chosen,

[1]Concerning the following notations known from physics and their relations to Geometric Algebra please refer to Chapt. 17 of [30]

for example, by Baylis [6]. According to the Born rule, the probability of the pure state $|0\rangle$ is $|a_0|^2$ and the probability of the pure state $|1\rangle$ is $|a_1|^2$. Finally, the total probability of the system is $\langle\psi|\psi\rangle = 1$, it follows the rule $|a_0|^2 + |a_1|^2 = 1$ and we can write $|\psi\rangle$ in spherical coordinates $(\gamma, \theta, \varphi)$ as an element of the three dimensional Bloch sphere

$$|\psi\rangle = e^{i\gamma}\left(\cos\left(\frac{\theta}{2}\right)|0\rangle + e^{i\varphi}\sin\left(\frac{\theta}{2}\right)|1\rangle\right),$$

which is equivalent to a group of unitary quaternions.

Thus, a quantum bit can be modelled in a complex space with two basis vectors (see equation (16.4)) or else in a real space with four basis vectors (see equation (16.6)). This second alternative can be found, e.g., in Doran and Lasenby (Chapt. 9 of [16]) or Cafaro and Mancini [8].

Our alternative representation of quantum bits can be motivated if one admits time-like vectors in addition to space-like ones, rather than using complex probability amplitudes. Using

$$a_0 = c_0^x + ic_0^t \text{ and } a_1 = c_1^x + ic_1^t, \tag{16.5}$$

the coefficients are transformed into real pre-factors thus doubling the number of basis vectors. In this alternative space-time notation of Geometric Algebra, the state vector of formula (16.4) takes the form:

$$\begin{aligned} \psi &= (c_0^x + ic_0^t)\sigma_0 + (c_1^x + ic_1^t)\sigma_1 \\ &= c_0^x\delta_0^x + c_0^t\delta_0^t + c_1^x\delta_1^x + c_1^t\delta_1^t, \end{aligned} \tag{16.6}$$

where we define $\delta_p^x = \sigma_p$ and $\delta_p^t = i\sigma_p$, $p \in \{0, 1\}$. The imaginary units can be taken over by the now time-like basis vectors if the measurements are performed with respect to only one space-like direction as with the quantum bits of a quantum computer. Equations (16.3) then change to:

$$\begin{aligned} (\delta_0^x)^2 = (\delta_1^x)^2 = 1; \\ (\delta_0^t)^2 = (\delta_1^t)^2 = -1. \end{aligned} \tag{16.7}$$

Please notice that these four elements cannot form an orthonormal basis of Clifford algebra $\mathscr{C}l(2,2)$, because of their non-anticommutativity, for example $\delta_0^x\delta_1^x = \sigma_0^x i\sigma_0^x = \delta_1^x\delta_0^x$. It is not even a commutative algebra because of the identity $\delta_0^x\delta_0^t = -\sigma_0^t\sigma_0^x = \delta_1^x\delta_0^x$.

We propose to model a qubit by unit elements of a Clifford algebra $\mathscr{C}l(2,2)$ which is a Clifford algebra over a vector space with an indefinite bilinear form of signature $(+,-,+,-)$. We identify this algebra with the vectors $(\gamma_0^x, \gamma_0^t, \gamma_1^x, \gamma_1^t)$, where

$$\begin{aligned} (\gamma_0^x)^2 = (\gamma_1^x)^2 = 1; \\ (\gamma_0^t)^2 = (\gamma_1^t)^2 = -1, \end{aligned} \tag{16.8}$$

and a qubit can be defined in correspondence with (16.6) as

$$\psi = c_0^x\gamma_0^x + c_0^t\gamma_0^t + c_1^x\gamma_1^x + c_1^t\gamma_1^t, \tag{16.9}$$

implying that qubits belong to a 3-dimensional surface defined by the identity

$$(c_0^x)^2 - (c_0^t)^2 + (c_1^x)^2 - (c_1^t)^2 = 1,$$

or to using standard hyperboloid coordinates $(\gamma, \theta, \varphi)$ equivalently,

$$|\psi\rangle = \bar{e}^{i\gamma}\left(\cos\left(\frac{\theta}{2}\right)|0\rangle + \bar{e}^{i\varphi}\sin\left(\frac{\theta}{2}\right)|1\rangle\right),$$

where $\bar{e}^{i\varphi} = \cosh(\varphi) + \sinh(\varphi)i$ and $\bar{e}^{i\gamma} = \cosh(\gamma) + \sinh(\gamma)i$.

16.3 QUANTUM REGISTER

A system consisting of several quantum bits is referred to as a quantum register. The simplest case is limited to two quantum bits. Figure 16.1 illustrates a space in which the corresponding wave function

$$|\Psi\rangle = a_{00}|00\rangle + a_{01}|01\rangle + a_{10}|10\rangle + a_{11}|11\rangle, \tag{16.10}$$

where $a_{ij} \in \mathbb{C}$ acts mathematically. This is actually an attempt to represent a four-dimensional space, with all four directions of the basic states $|00\rangle, |01\rangle, |10\rangle$, and $|11\rangle$ are pair-wise perpendicular.

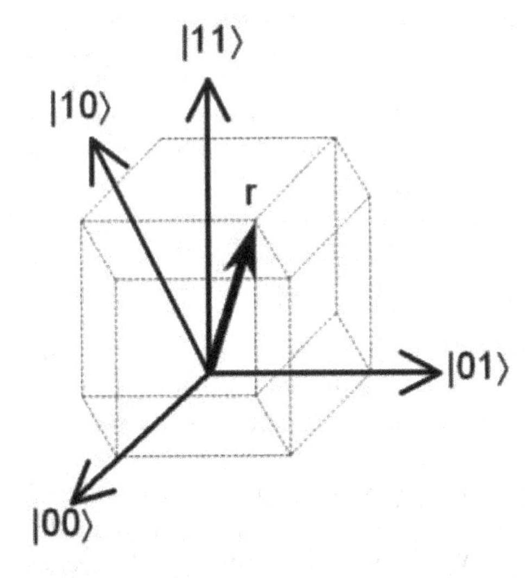

Figure 16.1 State vector ψ of a quantum register built from two quantum bits according to [41].

Again, there are two options available of Geometric Algebra representations. We can describe Ψ as an element of complex geometric algebra $\otimes^2 Cl(2, \mathbb{C})$ in the form

$$\Psi = a_{00}\sigma_{00} + a_{01}\sigma_{01} + a_{10}\sigma_{10} + a_{11}\sigma_{11}, \tag{16.11}$$

where $a_{ij} \in \mathbb{C}$ or as an element of real geometric algebra in the form

$$\Psi = c_{00}^x \gamma_{00}^x + c_{01}^x \gamma_{01}^x + c_{10}^x \gamma_{10}^x + c_{11}^x \gamma_{11}^x \\ + c_{00}^t \gamma_{00}^t + c_{01}^t \gamma_{01}^t + c_{10}^t \gamma_{10}^t + c_{11}^t \gamma_{11}^t, \tag{16.12}$$

where $\gamma_{jk}^i \in \otimes^2 Cl(2,2)$ and $a_{ij} \in \mathbb{R}$, so that the probability amplitudes are only real.

16.4 COMPUTING STEPS IN QUANTUM COMPUTING

Each computing step in a quantum computer actually transforms the wave function $|y\rangle$ or the state vector r that represents it. The following examples will show how the algebraic description of the transformation in the image of the wave function $|y\rangle$ can be supplemented by a geometric description that uses Geometric Algebra. The algebraic transformations are represented by reflections or rotations of the state vector r. This interplay between the algebraic and geometric representations within Geometric Algebra, enables a better grasp of quantum computing, which can thus be accessed in two different ways.

16.4.1 The NOT-Operation

One of the simplest operations is the simple inversion of the state of a single quantum bit. It is referred to as a NOT-operation. There are three different approaches to describing this operation:

Physical approach
Both states $|0\rangle$ and $|1\rangle$ are mapped on each other:

$$|0\rangle \rightarrow |1\rangle \textbf{ and } |1\rangle \rightarrow |0\rangle; \tag{16.13}$$

Algebraic approach
The two probability amplitudes are interchanged:

$$a_0 \rightarrow a_1 \textbf{ and } a_1 \rightarrow a_0; \tag{16.14}$$

Geometric approach
The state vector ψ is reflected about the reflection vector

$$r_{\text{ref}} = \frac{1}{\sqrt{2}}(\sigma_0 + \sigma_1), \tag{16.15}$$

See Fig. 16.2 for the geometric meaning of the reflection.

The actual physical implementation is achieved by irradiating the quantum bit with a suitable electromagnetic wave. The algebraic transformation is implemented by the transformer

$$U_{\text{NOT}} = |0\rangle\langle 1| + |1\rangle\langle 0| \tag{16.16}$$

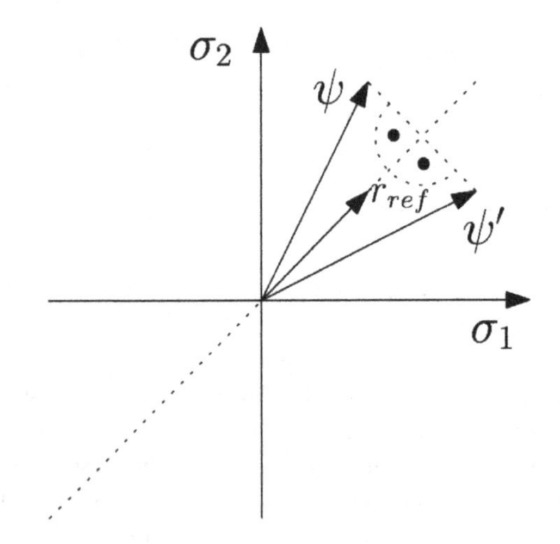

Figure 16.2 Reflection about the reflection vector.

as

$$\begin{aligned}\psi' &= U_{\text{NOT}}\,\psi \\ &= (|0\rangle\langle 1| + |1\rangle\langle 0|)(a_0|0\rangle + a_1|1\rangle) \\ &= a_1|0\rangle + a_0|1\rangle. \end{aligned} \tag{16.17}$$

In Geometric Algebra, using Pauli matrices, this transformation can now be written as the sandwich product

$$\begin{aligned}\psi' &= r_{\text{ref}}\,\psi\,r_{\text{ref}} \\ &= \frac{1}{\sqrt{2}}(\sigma_0+\sigma_1)(a_0\sigma_0+a_1\sigma_1)\frac{1}{\sqrt{2}}(\sigma_0+\sigma_1) \\ &= \frac{1}{2}(a_0\sigma_0^2 + a_0\sigma_1\sigma_0 + a_1\sigma_0\sigma_1 + a_1\sigma_1^2)(\sigma_0+\sigma_1) \\ &= \frac{1}{2}(a_0 + a_0\sigma_1\sigma_0 + a_1\sigma_0\sigma_1 + a_1)(\sigma_0+\sigma_1) \\ &= \frac{1}{2}(a_0 + a_0\sigma_1\sigma_0 + a_1\sigma_0\sigma_1 + a_1)(\sigma_0+\sigma_1) \\ &= a_1\sigma_0 + a_0\sigma_1 \end{aligned} \tag{16.18}$$

and, as such, handled as a reflection in the geometric sense.

An analogous geometric interpretation is obtained by using space-time basis vectors. This, however, requires two reflections. First a reflection is performed about the three-dimensional hyperplane

$$r_{\text{ref1}} = \frac{1}{\sqrt{2}}(\gamma_0^x + \gamma_1^x)\gamma_0^t\gamma_1^t, \tag{16.19}$$

to reflect the space-like basis vectors. Subsequently, the time-like vectors are reflected at the hyperplane

$$r_{\text{ref2}} = \frac{1}{\sqrt{2}}(\gamma_0^t + \gamma_1^t)\gamma_0^x\gamma_1^x, \tag{16.20}$$

resulting in

$$\psi' = r_{\text{ref2}} r_{\text{ref1}}\ \psi\ r_{\text{ref1}} r_{\text{ref2}} = -r_{\text{ref1}} r_{\text{ref2}}\ \psi\ r_{\text{ref1}} r_{\text{ref2}}. \tag{16.21}$$

As the composition of two reflections is a rotation, equation (16.21) can be thought of as a rotation. Let us point that, due to the anticommutative interchanging of different basis vectors:

$$\begin{aligned} r_{\text{ref1}} r_{\text{ref2}} &= \frac{1}{\sqrt{2}}(\gamma_0^x + \gamma_1^x)\gamma_0^t\gamma_1^t \frac{1}{\sqrt{2}}(\gamma_0^t + \gamma_1^t)\gamma_0^x\gamma_1^x \\ &= \frac{1}{2}(\gamma_0^x - \gamma_1^x)(\gamma_0^t - \gamma_1^t) \\ &= -r_{\text{ref2}} r_{\text{ref1}}. \end{aligned} \tag{16.22}$$

So, the element $r_{\text{ref1}} r_{\text{ref2}}$ represents a rotation, by conjugation with sign.

Thus, we have a rotation in the two-dimensional plane represented by the bivector

$$R = \frac{1}{2}(\gamma_0^x - \gamma_1^x)(\gamma_0^t - \gamma_1^t), \tag{16.23}$$

confirming the complete calculation:

$$\psi' = R\psi R = c_1^x\gamma_0^x + c_1^t\gamma_0^t + c_0^x\gamma_1^x + c_0^t\gamma_1^t. \tag{16.24}$$

Although this space-time description seems to demand much more computation power than the version of Geometric Algebra given by equation (16.18) working only with the space-like basis vectors, it is more instructive as the representations of reflections and rotations in spaces of higher dimensions become much clearer both algebraically and geometrically.

16.4.2 The Hadamard Transform

The Hadamard transform on a quantum bit is described by the matrix

$$H = \frac{1}{\sqrt{2}}\begin{pmatrix} 1 & 1 \\ 1 & -1 \end{pmatrix}. \tag{16.25}$$

In the higher-dimensional generalization it is very important for programming the entangled states of several quantum bits. Algebraically, it can be thought of as the transformation

$$U_H = \frac{1}{\sqrt{2}}(|0\rangle\langle 0| + |0\rangle\langle 1| + |1\rangle\langle 0| - |1\rangle\langle 1|) \tag{16.26}$$

with

$$r_H = U_H\, r = \frac{1}{\sqrt{2}}(a_0 + a_1)|0\rangle + \frac{1}{\sqrt{2}}(a_0 - a_1)|1\rangle. \tag{16.27}$$

Thus, the probability amplitudes are added or subtracted. In the Geometric Algebra context, we can see that this is actually a reflection about an axis which is rotated by $22.5^o = \pi/8$ about the state axis $|0\rangle$ (see, Fig. 16.3). Using the reflection vector

$$\begin{aligned} n &= \cos 22.5^o \sigma_0 + \sin 22.5^o \sigma_1 \\ &= \frac{1}{2}\sqrt{2+\sqrt{2}}\sigma_0 + \frac{1}{2}\sqrt{2-\sqrt{2}}\sigma_1, \end{aligned} \tag{16.28}$$

this can be easily shown:

$$\begin{aligned} r_H &= n\, r\, n \\ = \frac{1}{4}\left(\sqrt{2+\sqrt{2}}\sigma_0 + \sqrt{2-\sqrt{2}}\sigma_1\right)&(a_0\sigma_0 + a_0\sigma_1) \\ \frac{1}{4}&\left(\sqrt{2+\sqrt{2}}\sigma_0 + \sqrt{2-\sqrt{2}}\sigma_1\right) \\ &= \frac{1}{\sqrt{2}}(a_0 + a_1)\sigma_0 + \frac{1}{\sqrt{2}}(a_0 - a_1)\sigma_1 \end{aligned} \tag{16.29}$$

The computation confirms the graphical interpretation, but it can also be performed using space-time basis vectors. Here, in much the same way as in part 16.4.1,

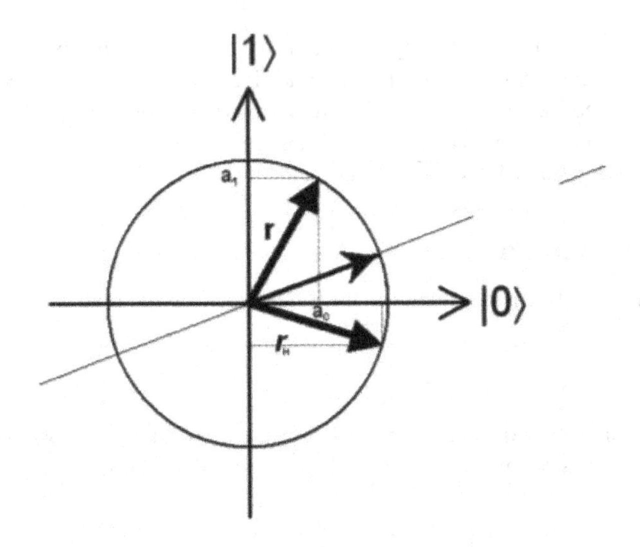

Figure 16.3 Graphical illustration of the Hadamard transform according to [41].

a double reflection is again applied to the state vector (16.6).

16.4.3 The CNOT-Operation

The CNOT-operation is one of the elementary logical operations that plays a crucial role acting on two qubits. If the first qubit has a value of 0 (appearing to be in the state $|0\rangle$), the state of the other qubit remains the same. If the first qubit has a value of 1 (appearing to be in the state $|1\rangle$), the other qubit will be inverted. In the usual notation, the algebraic transformation is implemented through the transformer

$$U_{CNOT} = |00\rangle\langle 00| + |01\rangle\langle 10| + |10\rangle\langle 11| + |11\rangle\langle 10| \tag{16.30}$$

as:

$$r_{CNOT} = U_{CNOT}\; r = a_{00}|00\rangle + a_{01}|01\rangle + a_{11}|10\rangle + a_{10}|11\rangle. \tag{16.31}$$

The effect is, that the coefficients a_{10} and a_{11} are interchanged.

In terms of Geometric Algebra, this transformation can be represented using the generalized Pauli matrices as a reflection of the state vector (16.11). This reflection in the four-dimensional state space takes place on a three-dimensional hyper plane. Firstly, this is generated by both the basis vectors not changed by the reflection representing the state directions. These directions are represented by the basis vectors σ_{00} and σ_{01}, since the first quantum bit has always the value 0. Secondly, a reflection is to be generated on the diagonals between the basis vectors σ_{10} and σ_{11} that interchanges both the directions. So the reflection plane looks like this:

$$m = \frac{1}{\sqrt{2}}\sigma_{00}\sigma_{01}(\sigma_{10} + \sigma_{11}) \tag{16.32}$$

In terms of Geometric Algebra, this transformation can now be written using Pauli matrices as

$$\begin{aligned} r_{CNOT} &= mrm && (16.33)\\ &= \frac{1}{\sqrt{2}}(\sigma_{00}\sigma_{01}(\sigma_{10} + \sigma_{11})\, r\, \frac{1}{\sqrt{2}}\sigma_{00}\sigma_{01}(\sigma_{10} + \sigma_{11}) && (16.34)\\ &= a_{00}\sigma_{00} + a_{01}\sigma_{01} + a_{11}\sigma_{10} + a_{10}\sigma_{11}. && (16.35) \end{aligned}$$

In much the same way, it is also easy to find a representation in Geometric Algebra with space-time basis vectors through the reflection of an eight-dimensional state vector (16.12) on the reflection plane

$$M = (\gamma^t_{10} - \gamma^t_{11})(\gamma^x_{10} - \gamma^x_{11}) \tag{16.36}$$

as:

$$\begin{aligned} r_{CNOT} &= MrM\\ &= c^x_{00}\gamma^x_{00} + c^x_{01}\gamma^x_{01} + c^x_{11}\gamma^x_{10} + c^x_{10}\gamma^x_{11}\\ &+ c^t_{00}\gamma^t_{00} + c^t_{01}\gamma^t_{01} + c^t_{11}\gamma^t_{10} + c^t_{10}\gamma^t_{11} \end{aligned} \tag{16.37}$$

CHAPTER 17

GAALOPWeb as a Qubit Calculator

CONTENTS

In this chapter, we use GAALOP (Geometric Algebra Algorithms Optimizer) to handle QBA (quantum bit algebra) as a Geometric Algebra for qubits according to [36]. We make all the calculations online via GAALOPWeb for Qubits. This enables us to make calculations for qubits without the need of installing a specific software. Since GAALOPWeb is webbrowser-based, it can be used on various devices such as PC, tablet or smart phone. We present how the NOT-operation can be applied on one as well as on two qubits.

We use Geometric Algebra for quantum computing motivated by the fact that qubits and gates can be handled as elements of the same algebra. Additionally, Geometric Algebra allows us to describe gate operations very easily, based on the geometrically intuitive description of transformations such as reflections or rotations in Geometric Algebra. The interplay between the algebraic and geometric representations within Geometric Algebra enables a better grasp of quantum computing. The goal is to have a mathematical language making the operations of quantum computing as intuitive as possible, in order to make the understanding of existing algorithms as well as the development of new algorithms as easy as possible. We describe quantum bits and quantum registers in a way mainly addressing the engineering community.

17.1 QUBIT ALGEBRA QBA

The Qubit Algebra is a 4D Geometric Algebra $G_{2,2}$ with one space-like basis vector and one time-like basis vector for each of the two basis states $|0\rangle$and $|1\rangle$. Table

DOI: 10.1201/9781003139003-17

17.1 shows the 4 basis vectors together with their relation to the qubit states, their signature and their description within GAALOP.

TABLE 17.1 The 4 Basis Vectors of the Qubit Algebra

	basis vector	Signature	GAALOPScript
$\|0\rangle$ (space-like)	γ_0^x	+1	e0x
$\|0\rangle$ (time-like)	γ_0^t	-1	e0t
$\|1\rangle$ (space-like)	γ_1^x	+1	e1x
$\|1\rangle$ (time-like)	γ_1^t	-1	e1t

17.2 GAALOPWeb FOR QUBITS

GAALOPWeb is our online tool for optimizing Geometric Algebra algorithms for programming languages. We adapted the recent interface in order to support Qubit algebras[1], see Fig. 17.1.

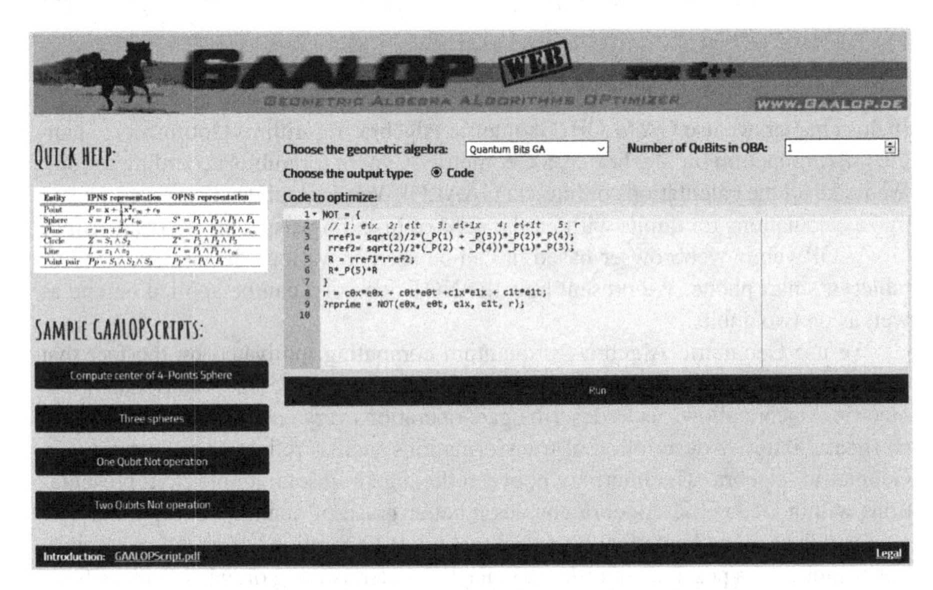

Figure 17.1 Screenshot of GAALOPWeb for Qubit algebras.

After the user selected the *Quantum Bits Geometric Algebra* as geometric algebra, a number selection field appears where the user is able to set the number of qubits. Then the user edits the code written in GAALOPScript and pushes the *Run* button. In the background, a new algebra definition is created automatically based on

[1] http://www.gaalop.de/gaalopweb

the number of qubits and this is used to compute the resulting C/C++ Code, which is displayed to the user afterwards.

17.3 THE NOT-OPERATION ON A QUBIT

Listing 17.1 shows the GAALOPScript for the NOT-operation according to section 16.4.1.

Listing 17.1: GAALOPScript for the NOT-operation on a qubit

```
1 r = c0x*e0x + c0t*e0t +c1x*e1x + c1t*e1t;
2 rref1= sqrt(2)/2*(e0x + e1x)*e0t*e1t;
3 rref2= sqrt(2)/2*(e0t + e1t)*e0x*e1x;
4 R = rref1*rref2;
5 ?rprime = R*r*R;
```

Line 1 computes the general state vector r as a linear combination of all the basis vectors according to equation (16.6). The line 2 and 3 are responsible for the computation of the hyper planes for the two needed reflections according to the equations (16.19) and (16.20). Line 4 computes the combined reflector R for both reflections according to equation (16.22) and finally the transformed state vector is computed as the sandwich product according to equation (16.24).

The result of the generation of C/C++ code is as follows:

Listing 17.2: Result of the NOT-operation

```
1 void calculate(float c0t, float c0x, float c1t, float c1x,
2             float rprime[16]) {
3
4       rprime[1] = c1x; // e0x
5       rprime[2] = c0x; // e1x
6       rprime[3] = c1t; // e0t
7       rprime[4] = c0t; // e1t
8 }
```

It shows the new coefficients of the basis vectors e0x, e1x, e0t and e1t. Compared to the original state vector r, we realize that both the coefficients of the two space-like basis vectors and of the two time-like basis vectors are now interchanged.

17.4 THE 2-QUBIT ALGEBRA QBA2

The following table shows the 8 basis vectors of the 2-Qubit Algebra QBA2 with their signatures and GAALOP notations.

In order to compute the NOT-operation on the first qubit, we are able to use the same GAALOPScript according to Listing 17.1 as used for QBA. The result as shown in Listing 17.3 is in principle the same as in Listing 17.2 (the only difference is an internal one: the indices of the data structure for the multivector rprime).

TABLE 17.2 The 8 Basis Vectors of the 2-Qubit Algebra QBA2

	basis vector	Signature	GAALOPScript
$\|00\rangle$ (space-like)	γ_{00}^{x}	+1	e0x
$\|00\rangle$ (time-like)	γ_{00}^{t}	-1	e0t
$\|01\rangle$ (space-like)	γ_{01}^{x}	+1	e1x
$\|01\rangle$ (time-like)	γ_{01}^{t}	-1	e1t
$\|10\rangle$ (space-like)	γ_{10}^{x}	+1	e2x
$\|10\rangle$ (time-like)	γ_{10}^{t}	-1	e2t
$\|11\rangle$ (space-like)	γ_{11}^{x}	+1	e3x
$\|11\rangle$ (time-like)	γ_{11}^{t}	-1	e3t

Listing 17.3: Result of the NOT-operation on the first qubit.

```
1  void calculate(float c0t, float c0x, float c1t, float c1x,
2                 float rprime[256]) {
3
4         rprime[1] = c1x; // e0x
5         rprime[2] = c0x; // e1x
6         rprime[5] = c1t; // e0t
7         rprime[6] = c0t; // e1t
8  }
```

If we would like to use the NOT-operation for the second qubit, we have to take the corresponding basis vectors according to Listing 17.4.

Listing 17.4: GAALOPScript for the NOT-operation on the second qubit in QBA2.

```
1  r = c2x*e2x + c2t*e2t +c3x*e3x + c3t*e3t;
2  rref1= sqrt(2)/2*(e2x + e3x)*e2t*e3t;
3  rref2= sqrt(2)/2*(e2t + e3t)*e2x*e3x;
4  R = rref1*rref2;
5  ?rprime = R*r*R;
```

We can see in the result of the Listing 17.5 that again both the coefficients of the two space-like basis vectors and of the two time-like basis vectors are now interchanged.

Listing 17.5: Result of the NOT-operation on the second qubit.

```
1
2  void calculate(float c2t, float c2x, float c3t, float c3x,
3                 float rprime[256]) {
4
5         rprime[3] = c3x; // e2x
6         rprime[4] = c2x; // e3x
7         rprime[7] = c3t; // e2t
8         rprime[8] = c2t; // e3t
9  }
```

In order to make code reusable, GAALOPScript offers a macro concept. This allows to write a macro for the NOT-operation which can be used for both qubits as shown in Listing 17.6.

Listing 17.6: NOT-operation on both qubits using a macro.

```
1  NOT = {
2   // 1: eix   2: eit    3: ei+1x    4: ei+1t
3   rref1= sqrt(2)/2*(_P(1) + _P(3))*_P(2)*_P(4);
4   rref2= sqrt(2)/2*(_P(2) + _P(4))*_P(1)*_P(3);
5   R = rref1*rref2;
6   R*_P(5)*R
7  }
8  r1 = c0x*e0x + c0t*e0t +c1x*e1x + c1t*e1t;
9  ?rprime1 = NOT(e0x, e0t, e1x, e1t, r1);
10
11 r2 = c2x*e2x + c2t*e2t +c3x*e3x + c3t*e3t;
12 ?rprime2 = NOT(e2x, e2t, e3x, e3t, r2);
```

The meaning of the parameters `_P(1)` ... `_P(5)` of the NOT macro is indicated in line 2. The result according to Listing 17.7 is now a summary of the results for each of the qubits.

Listing 17.7: Result of the NOT-operation on both qubits.

```
1  void calculate(float c0t, float c0x, float c1t, float c1x,
2                float c2t, float c2x, float c3t, float c3x,
3                float rprime1[256], float rprime2[256]) {
4
5          rprime1[1] = c1x; // e0x
6          rprime1[2] = c0x; // e1x
7          rprime1[5] = c1t; // e0t
8          rprime1[6] = c0t; // e1t
9          rprime2[3] = c3x; // e2x
10         rprime2[4] = c2x; // e3x
11         rprime2[7] = c3t; // e2t
12         rprime2[8] = c2t; // e3t
13 }
```

CHAPTER 18

Appendix

CONTENTS

18.1 APPENDIX A: PYTHON CODE FOR THE GENERATION OF OPTIMIZED MATHEMATICA CODE FROM GAALOP

```
def replaceFunctionCall (line, fold, fnew):
    # handle fold -> fnew correctly (especially specific brackets)
    # 'sqrtf' -> 'Sqrt'
    # 'fabs', 'Abs'

    SplitStr = line.split(fold + '(')
    if len(SplitStr) == 2:
        #print(fold + ' ->' + fnew)
        Rest = SplitStr[1]
        counter=1
        #print('Len: ' + str(len(Rest)) + '\n')
        # compute the index of the character to be changed
        # in the string Rest
        indexInRest=0
        for i in range(0, len(Rest)):
            if Rest[i] == '(':
                counter=counter+1
            if Rest[i] == ')':
                counter=counter-1
            if counter==0:
                #print('Wert: ' + str(i) + Rest[i] + '\n')

                indexInRest=i
                break
        line = SplitStr[0] + fnew + '['
        # add rest
        for i in range(0, len(Rest)):
            if i==indexInRest:
                line = line + ']'
```

DOI: 10.1201/9781003139003-18

```
        else:
            line = line + Rest[i]

    return line

def GAALOPtoMathematica (GAFunctionName, *mv):

    f=open(GAFunctionName + '.c', 'r')
    PythonFile = open(GAFunctionName + '.txt', 'w')

    Lines = f.readlines()
    length = len(Lines)

    line = Lines[0]
    line = line.replace(',', '_,') # add underscore to parameters
    line = line.replace('float', '')     # delete all 'float's

    # compute the length of the multivectors
    # looking for the first multivector length ...
    SplitStr = line.split('[')
    mv_length = SplitStr[1]
    mv_length = mv_length.split(']')[0]

    # compute list of all multivectors
    # split at the max length of the multivectors ...
    mvList = line.split('['+mv_length+']')
    # correct the first element
    SplitLst = mvList[0].split('_,  ')
    mvStr = SplitLst[len(SplitLst)-1]
    mvList[0] = mvStr
   # ignore last element of the list
    for i in range(1, len(mvList)-1):
        mvList[i] = mvList[i].replace('_,  ', '')
        mvStr += ', ' + mvList[i]

    # first line
    line = line.replace('void calculate(',  GAFunctionName + '[')
    SplitLst = line.split(',  ' + mvList[0])
    line = SplitLst[0]

    line = line + '] := Module[{' + mvStr + '},\n'
    PythonFile.write(line)

    # second lines with array declarations for multivectors
    for i in range(0, len(mvList)-1):
        SecondLine = mvList[i] + ' = ConstantArray[0,' + mv_length + '];\n'
        PythonFile.write(SecondLine)

    # generate lines for the coefficients of the multivectors
```

```
for i in range(1, length-1):
    line = Lines[i]s
    line = line.lstrip() # deletes empty lines
    if len(line) > 0:
        # replace characters for comments and specific functions
        line = line.replace('// ', '(*')
        line = line.replace('\r\n', '*)\n')
        line = line.replace('\r', '*)\n')
        line = line.replace('\n', '*)\n')
        #line = line.replace('fabs', 'Abs')

        # integrate the correct index (+1)
        SplitStr = line.split('[')
        line = SplitStr[0]
        for i in range(1,len(SplitStr)):
            # integrate the correct index on the left side
            IndexLst = SplitStr[i].split(']')
            Index = IndexLst[0]
            Index = str(int(Index)+1)          # add 1 to Index
            line = line + '[[' + Index + ']]' + IndexLst[1]

        # handle function calls correctly
        line = replaceFunctionCall (line, 'sqrtf', 'Sqrt')
        line = replaceFunctionCall (line, 'fabs', 'Abs')
        line = replaceFunctionCall (line, 'cos', 'Cos')
        line = replaceFunctionCall (line, 'sin', 'Sin')
        line = replaceFunctionCall (line, 'pow', 'Power')


        PythonFile.write(line)

# last line in Mathematica style
Strmvs = mv[0]
for i in range(1, len(mv)):
    Strmvs += ', ' + mv[i]
line = 'Return[{' + Strmvs + '}];]'
PythonFile.write(line)

PythonFile.close()
f.close()
```

Bibliography

[1] Ahmad Eid H. A. Gmac: Geometric macro homepage. Available at `https://gmac-guides.netlify.com`, 2018. Last visited June 2021

[2] Rafael Alves, Dietmar Hildenbrand, Christian Steinmetz, and Patrick Uftring. Efficient development of competitive mathematica solutions based on geometric algebra with gaalopweb. *Advances in Applied Clifford Algebras Journal*, 2020.

[3] Rafael Alves, Carlile Lavor, Cipriano de Souza, and Michael Souza. Clifford algebra and discretizable distance geometry. *Mathematical Methods in the Applied Sciences*, 41(11):4063–4073, 2017.

[4] Andreas Aristidou and Joan Lasenby. Fabrik. *Graphical Models*, 73(5):243–260, 2011.

[5] Alex Arsenovic, Hugo Hadfield, Eric Wieser, Robert Kern, and The Pygae Team. Available at https://zenodo.org/record/3874239#. YMI1_fkzY2w

[6] William Baylis. A relativistic algebraic approach to the qc interface: Implications for quantum reality. *Advances in Applied Clifford Algebras*, 2008.

[7] John Browne. The Grassmann algebra package home page. Available at `http://sites.google.com/site/grassmannalgebra/`, 2009. Last visited June 6, 2021

[8] Carlo Cafaro and Stefano Mancini. A geometric algebra perspective on quantum computational gates and universatility in quantum computing. *Advances in Applied Clifford Algebras*, 2011.

[9] V.S. Camargo, Castelani E.V., L.A.F. Fernandes, and F Fidalgo. Geometric algebra to describe the exact discretizable molecular distance geometry problem for an arbitrary dimension. *Advances in Applied Clifford Algebras*, 2019.

[10] Andrea Cassioli, Benjamin Bardiaux, Guillaume Bouvier, Antonio Mucherino, Rafael Alves, Leo Liberti, Michael Nilges, Carlile Lavor, and Therese E. Malliavin. An algorithm to enumerate all possible protein conformations verifying a set of distance constraints. *BMC Bioinformatics*, 16(1):23, 2015.

[11] William Kingdon Clifford. *Applications of Grassmann's Extensive Algebra*, volume 1 of *American Journal of Mathematics*, pages 350–358. The Johns Hopkins University Press, 1878.

[12] Pablo Colapinto. The versor home page. Available at http://versor.mat.ucsb.edu/, 2015. Last visited June 6, 2021

[13] Steven De Keninck. PGA Cheat sheets, available at www.bivector.net, 2020.

[14] Steven De Keninck. ganja.js, available at https://github.com/enkimute/ganja.js, 2020.

[15] Steven De Keninck. ganja.js coffeeshop, 2020.

[16] Chris Doran and Anthony Lasenby. *Geometric Algebra for Physicists*. Cambridge University Press, 2003.

[17] Leo Dorst. Boolean combination of circular arcs using orthogonal spheres. *Advances in Applied Clifford Algebras*, 2019.

[18] Robert Benjamin Easter and Eckhard Hitzer. Double conformal geometric algebra. *Advances in Applied Clifford Algebras Journal*, 2017.

[19] Robert Benjamin Easter and Eckhard Hitzer. Conic and cyclidic sections in double conformal geometric algebra g8,2 with computing and visualization using gaalop. *Mathematical Methods in the Applied Sciences*, 2019.

[20] Ahmad Hosney Awad Eid. *Optimized Automatic Code Generation for Geometric Algebra Based Algorithms with Ray Tracing Application.* PhD thesis, Suez Canal University, Port Said, 2010.

[21] Daniel Fontijne, Tim Bouma, and Leo Dorst. Gaigen 2: A geometric algebra implementation generator. Available at `http://staff.science.uva.nl/~fontijne/gaigen2.html`, 2007.

[22] S. Franchini, A. Gentile, F. Sorbello, G. Vassallo, and S. Vitabile. Conformalalu: A conformal geometric algebra coprocessor for medical image processing. *IEEE Transactions on Computers*, 64(4):955–970, April 2015.

[23] Luis Enrique Gonzalez-Jimenez, Oscar Eleno Carbajal-Espinosa, and Eduardo Bayro-Corrochano. Geometric techniques for the kinematic modeling and control of robotic manipulators. In *2011 IEEE International Conference on Robotics and Automation*, pages 5831–5836. IEEE, 2011.

[24] Hermann Grassmann. *Die Ausdehnungslehre. Vollstaendig und in strenger Form begruendet.* Verlag von Th. Chr. Fr. Enslin, Berlin, 1862.

[25] Hugo Hadfield, Dietmar Hildenbrand, and Alex Arsenovic. Gajit: Symbolic optimisation and jit compilation of geometric algebra in python with gaalop and numba. *Proceedings of CGI conference Calgary, Canada*, 2019.

[26] David Hestenes. Old wine in new bottles: A new algebraic framework for computational geometry. In Eduardo Bayro-Corrochano and Garret Sobczyk, editors, *Geometric Algebra with Applications in Science and Engineering*. Birkhäuser, 2001.

[27] David Hestenes. The genesis of geometric algebra – a personal retrospective. In *Advances in Applied Clifford Algebras Journal*, open access at Springerlink.com, 2016.

[28] Dietmar Hildenbrand. *Geometric Computing in Computer Graphics and Robotics using Conformal Geometric Algebra*. PhD thesis, TU Darmstadt, 2006. Darmstadt University of Technology.

[29] Dietmar Hildenbrand. *Foundations of Geometric Algebra Computing*. Springer, 2013.

[30] Dietmar Hildenbrand. *Introduction to Geometric Algebra Computing*. Taylor & Francis Group, 2019.

[31] Dietmar Hildenbrand, Justin Albert, Patrick Charrier, and Christian Steinmetz. Geometric algebra computing for heterogeneous systems. *Advances in Applied Clifford Algebras Journal*, 2016.

[32] Dietmar Hildenbrand, Patrick Charrier, Christian Steinmetz, and Joachim Pitt. GAALOP home page. Available at `http://www.gaalop.de`, 2020.

[33] Dietmar Hildenbrand, Patrick Charrier, Christian Steinmetz, and Joachim Pitt. GAALOP home page. Available at `http://www.gaalop.de`, 2020.

[34] Dietmar Hildenbrand, Daniel Fontijne, Yusheng Wang, Marc Alexa, and Leo Dorst. Competitive runtime performance for inverse kinematics algorithms using conformal geometric algebra. In *Eurographics Conference Vienna*, 2006.

[35] Dietmar Hildenbrand, Silvia Franchini, A. Gentile, G. Vassallo, and S. Vitabile. Gappco: an easy to configure geometric algebra coprocessor based on gapp programs. *Advances in Applied Clifford Algebras Journal*, 2017.

[36] Dietmar Hildenbrand, Christian Steinmetz, Rafael Alves, Jaroslav Hrdina, and Carlile Lavor. An online calculator for qubits based on geometric algebra. *Proceedings of CGI conference Geneva, Switzerland*, 2020.

[37] Dietmar Hildenbrand, Christian Steinmetz, and Radek Tichy. Gaalopweb for matlab: an easy to handle solution for industrial geometric algebra implementations. *Advances in Applied Clifford Algebras Journal*, 2020.

[38] Eckhard Hitzer and Dietmar Hildenbrand. Cubic curves and cubic surfaces from contact points in conformal geometric algebra. *Proceedings of CGI conference Calgary, Canada*, 2019.

[39] Eckhard Hitzer and Steve Sangwine. Multivector and multivector matrix inverses in real Clifford algebras. Technical report, University of Essex, 2016.

[40] Matthias Homeister. *Quantum Computing verstehen. Grundlagen - Anwendungen - Perspektiven.* Friedrich Vieweg & Sohn Verlag, 2018.

[41] Martin Erik Horn, Paul Drechsel, and Dietmar Hildenbrand. Quanten-computing und geometrische algebra. *Proceeedings Didaktik der Physik*, 2012.

[42] J. Hrdina, A. Navrat, and P. Vasik. Notes on planar inverse kinematics based on geometric algebra. *Advances in Applied Clifford Algebras*, 28(71):1–14, 2018.

[43] Jaroslav Hrdina, Ales Navrat, and Petr Vask. Geometric algebra for conics. *Advances in Applied Clifford Algebras Journal*, 2018.

[44] George Johnson. *A shortcut through time. The path to the quantum computer.* Vintage books New York, 2004.

[45] Mitsuhiro Kondo, Takuya Matsuo, Yoshihiro Mizoguchi, and Hiroyuki Ochiai. A mathematica module for conformal geometric algebra and origami folding. In James H. Davenport and Fadoua Ghourabi, editors, *SCSS 2016. 7th International Symposium on Symbolic Computation in Software Science*, volume 39 of *EPiC Series in Computing*, pages 68–80. EasyChair, 2016.

[46] C. Lavor, L. Liberti, N. Maculan, and A. Mucherino. The discretizable molecular distance geometry problem. *Computational Optimization and Applications*, 52:115–146, 2012.

[47] Carlile Lavor and Rafael Alves. Oriented conformal geometric algebra and the molecular distance geometry problem. *Advances in Applied Clifford Algebras*, 29(9), 2019.

[48] Carlile Lavor, Rafael Alves, Weber Figueiredo, Antonio Petraglia, and Nelson Maculan. Clifford algebra and the discretizable molecular distance geometry problem. *Advances in Applied Clifford Algebras*, 25:925–942, 2015.

[49] "Carlile Lavor, Leo Liberti, Bruce Donald, Bradley Worley, Benjamin Bardiaux, Therese E. Malliavin, and Michael Nilges. Minimal nmr distance information for rigidity of protein graphs. *Discrete Applied Mathematics*, 256:91–104, 2019.

[50] Hongbo Li, David Hestenes, and Alyn Rockwood. Generalized homogeneous coordinates for computational geometry. In G. Sommer, editor, *Geometric Computing with Clifford Algebra*, pages 27–59. Springer, 2001.

[51] Leo Liberti, Carlile Lavor, and Nelson Maculan. A branch-and-prune algorithm for the molecular distance geometry problem. *International Transactions in Operational Research*, 15(1):1–17, 2008.

[52] Leo Liberti, Carlile Lavor, Nelson Maculan, and Antonio Mucherino. Euclidean distance geometry and applications. *SIAM Review*, 56(1):3–69, 2014.

[53] Stephen Mann, Leo Dorst, and Tim Bouma. The making of GABLE, a geometric algebra learning environment in Matlab. pages 491–511, 2001.

[54] Maxima Development Team. Jupyter homepage. Available at `https://jupyter.org/`, 2017.

[55] David McMahon. *Quantum Computing Explained*. John Wiley & Sons, 2008.

[56] A. Mucherino and C. Lavor. The branch and prune algorithm for the molecular distance geometry problem with inexact distances. In *Proceedings of the International Conference on Computational Biology*, 2009.

[57] Christian Perwass. *Geometric Algebra with Applications in Engineering*. Springer, 2009.

[58] Project Jupyter. Maxima, a computer algebra system. version 5.18.1. Available at `http://maxima.sourceforge.net/`, 2020.

[59] Stephen J. Sangwine and Eckhard Hitzer. Clifford multivector toolbox (for matlab). *Advances in Applied Clifford Algebras*, 27(1):539–558, 2017.

[60] Florian Seybold. Gaalet – a C++ expression template library for implementing geometric algebra. Technical report, 2010.

[61] Michael Souza. Mdgp package. Available at `https://github.com/michaelsouza/bioinfo/tree/master/codes/IMPA2017`, 2017.

[62] Christian Steinmetz. Optimizing a geometric algebra compiler for parallel architectures using a table-based approach. In *Bachelor thesis TU Darmstadt*, 2011.

[63] Christian Steinmetz. Examination of new geometric algebras including a visualization and integration in a geometric algebra compiler. Master's thesis, 2013.

[64] Florian Stock, Dietmar Hildenbrand, and Andreas Koch. Fpga-accelerated color edge detection using a geometric-algebra-to-verilog compiler. In *Symposium on System on Chip (SoC)*, 2013.

[65] Terje Vold. Geometric algebra library. Available at `https://library.wolfram.com/infocenter/Conferences/6951/`, 2007.

[66] Florian Woersdoerfer, Florian Stock, Eduardo Bayro-Corrochano, and Dietmar Hildenbrand. Optimization and performance of a robotics grasping algorithm described in geometric algebra. In *Iberoamerican Congress on Pattern Recognition 2009, Guadalajara, Mexico*, 2009.

[67] Stephen Wolfram. *The Mathematica Book*. Wolfram Media, Incorporated, 5 edition, 2003.

[68] Julio Zamora-Esquivel. G6,3 geometric algebra. In *ICCA9, 7th International Conference on Clifford Algebras and Their Applications*, 2011.

Index

Printed in the United States
by Baker & Taylor Publisher Services

Printed in the United States
by Baker & Taylor Publisher Services